revision guides

Do**Brilliantly**

ASGeography

Exam practice at its **best**

- **Michael Raw**
- **Series Editor: Jayne de Courcy**

Contents

Published by HarperCollins*Publishers* Limited
77–85 Fulham Palace Road
London W6 8JB

www.**Collins**Education.com
On-line support for schools and colleges

First published 2002
10 9 8 7 6 5 4 3 2
ISBN 0 00 712430 9

Michael Raw asserts the moral right to be identified as the author of this work.

British Library Cataloguing in Publication Data
A catalogue record for this book is available from the British Library.

Edited by Joan Miller
Production by Kathryn Botterill
Cover design by Susi Martin-Taylor
Book design by Gecko Limited
Printed and bound in China by Imago

Acknowledgements
The Author and Publishers are grateful to NERC, University of Dundee, for permission to reproduce the photograph on page 32.

Illustrations
Roger Penwill, Gecko Limited

You might also like to visit:
www.**fire**and**water**.com
The book lover's website

How this book will help you
by Michael Raw

This book will help you improve your performance in your AS Geography exam.

Exam practice – how to answer questions better

To succeed at AS level you need two things:
- good knowledge and understanding of the geographical content required by your exam board's specification
- effective examination technique which allows you to apply your knowledge and understanding appropriately to the questions.

It is the second of these that this book will help you improve.

Each chapter in the book is broken down into three separate elements:

1 Exam question, sample answer and 'How to score full marks'

The **questions and answers are typical**, showing most of the mistakes made by candidates under exam conditions.

'How to score full marks' shows where the sample answers would lose marks, for example, by not obeying command words or not putting in enough detailed geographical information. **I show you how to gain extra marks** so that when you meet questions of this type in your exam you will know exactly how to answer them.

2 Don't forget

A book of this size can't include all the geographical information that you need to know. However, in the **'Don't forget'** boxes, I have listed some key topics that you must revise in detail from your notes or a textbook. **These boxes also summarise important aspects of exam technique.**

3 Questions to try, Answers and Examiner's hints

Each chapter ends with an exam question for you to attempt. Don't cheat. Sit down and try to answer the question as if you were in an exam. Try to remember all that you've read earlier in the chapter and put it into practice.

Check your answer then look at the answer given at the back of the book. These answers would gain full marks. The **'What makes this a good answer?' section highlights where marks are gained**. Compare your answers with those given and decide whether your answers would gain full marks and, if not, what aspects you need to improve.

AS level specifications

There are seven AS level Geography specifications offered by examining boards in England and Wales. Their most striking feature is their similarity.

They all cover physical geography, human geography, the relationships between the physical environment and human activities, and geographical skills.

They all have written examinations, based on short-answer and extended-answer structured questions, lasting either 1 hour 15 minutes or 1 hour 30 minutes.

However, each AS specification has its own distinctive content and emphasis. Ensure that you know in detail the content of *your own* specification and use the relevant chapters in this book.

Structured questions

The first thing to establish is the type of question used for your specification. All specifications use structured questions: most are short-answer questions, requiring responses of two to five lines. However, within a question there are often parts that require longer answers. These extended answer questions, which often demand case studies and place specific examples, usually require answers of 20 to 30 lines.

This book contains examples of structured questions for all of the examining bodies.

Short-answer questions

Short-answer structured questions have most if not all of the following features:

- several sub-questions requiring answers which typically range from one or two lines to seven or eight lines
- questions usually built around one or two stimulus materials such as sketch maps, OS maps, diagrams, charts, photographs and tables of data
- a gradient of difficulty, starting with simple questions of fact or definition, followed by explanation and sometimes extended answer sub-questions requiring the use of examples or case studies.

The simplest questions may ask for information derived directly from the stimulus materials. However, most sub-questions are stimulus-response. This means that the answers are not found in the stimulus materials themselves. Rather these questions invite you to apply your knowledge and critical understanding to a specific problem or situation that you have not come across before. Look at the sample question on beach particles below and my comments on the types of questions.

	Number of particles	
	Beach A (north-facing)	**Beach B (south-facing)**
Very angular	77	1
Angular	88	11
Sub-angular	31	99
Sub-rounded	19	22
Rounded	23	31
Well rounded	12	36

Table 1 Roundness of limestone shingle at Arnside, north Lancashire

a) Describe the differences in particle size distribution for the two beaches in Table 1. (2)

b) State the modal size of particles on beaches A and B. (2)

c) Name and justify the use of one type of chart to represent the data in Table 1. (4)

d) Outline one process responsible for the rounding of beach particles. (4)

e) Using the evidence of Figure 1, suggest a possible explanation for differences in particle roundness on the two beaches. (6)

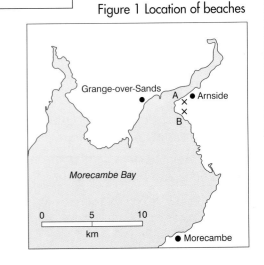

Figure 1 Location of beaches

Question types

a and b: data-response questions

These questions are data-response type. Question a requires an accurate description of the frequency distribution of particles on the two beaches. No additional knowledge or understanding is needed. In question b, provided you understand the term mode, you can find the correct answer from the data. You should note that these questions, which make fewer demands than c, d and e, have a lower mark weighting.

c, d and e: stimulus-response questions

These are stimulus-response questions. Table 1 does not contain the answers to these questions. An effective response to question c depends on your knowledge and understanding of histograms. In question d you need to know how wave action, through the processes of abrasion and attrition, rounds beach particles. Question e demands an understanding of how differences in the aspect of coasts determine the fetch and therefore the amount of wave energy input on a beach. In this example, beach B has a longer fetch than beach A, therefore beach B should receive greater inputs of wave energy than beach A. The greater erosive energy at B gives more rounded sediments. These questions are worth 4–6 marks each.

Extended-answer questions

Most AS specifications give some scope for extended answers to structured questions. These extended-answer questions are usually the final (and most demanding) part of a short-answer structured question.

Three examples are given below.

1 Referring to examples, explain how beach profiles are influenced by the physical processes that operate on them.   (10)

2 Out-of-centre developments for retailing and leisure threaten the status of the CBD. Outline, with examples, the arguments for these out-of-centre developments.   (10)

3 With reference to one or more urban areas in the UK, explain how social and economic changes over the past 30 years have affected the type and location of housing developments.   (10)

How structured questions are marked

Examiners have two strategies for marking structured questions. First, questions worth 2–3 marks are always **point-marked**. There may be a mark for stating a valid idea or reason, and two marks for development or quality of explanation. Second, levels of response mark schemes are often used for short-answer questions worth 4–6 marks. **Level-marking** is always used for extended-answer questions.

In levels of response mark schemes, the quality of your answer will be assessed against criteria such as accuracy and appropriateness of knowledge and understanding, factual detail and use of examples. A possible mark scheme for the three extended-answer questions in the preceding section is given below.

Level 1: 0–3 marks. Basic knowledge and understanding which has some relevance to the question.

Level 2: 4–6 marks. Clear and accurate knowledge and understanding which is largely appropriate to the question.

Level 3: 7–10 marks. Detailed and accurate knowledge and understanding, wholly appropriate to the question.

Tackling structured questions

- Study the stimulus material carefully. Make sure that you have a thorough understanding of the information it contains.
- Read through all parts of the question before attempting to answer. Each question should be a coherent unit: the sub-parts should link together and show logical development and there should be a gradient of difficulty. Reading through the whole question should help avoid repetition in your answers.
- Take special care with the command words and phrases in each question (see page 6). Underline these key words and phrases and be sure to distinguish between commands requiring description and those requiring explanation.
- Plan your answers before writing. You have a limited space to answer each question and may be penalised for writing excessively. Be precise and economical in your use of language.
- Where appropriate, include brief examples. Questions which include definitions may give scope for the appropriate use of an example (even if not specifically asked for); in a marginal answer they may just add enough to tip the balance in your favour.

Common errors in answering structured questions

Question 1	What is meant by the term 'stream discharge'?
Answer	Stream discharge is the flow of water in a stream channel.
Error	Inaccurate/insufficient knowledge and understanding.
Examiner's comment	The answer lacks precision and accuracy, and is too generalised. The candidate's definition could apply equally to the velocity of flowing water in a channel. The answer gains no marks. The correct answer is that discharge describes the volume of flow over a given period of time. It is usually measured in cubic metres per second (cumecs). This answer as it stands is not creditable. A reference to cumecs would have secured some marks.
Question 2	State and explain two factors which influence stream discharge.
Answer 1	1 The size of the area (i.e. catchment) drained by a river. 2 The number of tributaries that join the river.
Error 1	Repetition.
Examiner's comment	Both answers focus on the same factor i.e. size of the catchment. The number of tributaries joining the river merely reflects the size of the catchment. The second answer should provide a factor that is different from the first e.g. vegetation cover of the catchment. The first answer would be credited with full marks; the second gains no credit.
Answer 2	1 Discharge is highest in winter when interception is low. In winter most trees have no foliage and ground vegetation dies down. 2 Discharge is lowest in summer when interception is high. At this time of year trees are in foliage and ground vegetation is fully grown.
Error 2	Mirror answer: one answer is the reverse of the other.
Examiner's comment	Both answers deal with the same factor: interception of precipitation by vegetation surfaces. The examiner will credit only one answer.
Answer 3	1 Amount of evapotranspiration. 2 Amount of soil moisture.
Error 3	Failure to explain or make connections.
Examiner's comments	Both answers are essentially correct but the candidate merely states the factors and does not explain the causal linkages between stream discharge, evapotranspiration and soil moisture. The candidate assumes that the answers are self evident. They are not, and the onus is on the candidate to make the connections explicit. No credit can be given for the second part of the question.
Answer 4	Stream discharge occurs when water runs-off the surface in channels. The water enters stream channels via overland flow, through flow, groundwater flow and directly through channel precipitation.
Error 4	Failure to respond accurately to command words.
Examiner's comments	The answer does not address the question. The candidate shows knowledge of the movement of water into stream channels, but fails to apply knowledge appropriately to the question. No credit can therefore be given for this answer.
Answer 5	1 Type of precipitation (e.g. rain or snow) influences discharge. 2 Geology (e.g. permeable or impermeable rocks) influence discharge.
Error 5	Failure to develop answers.
Examiner's comments	Excessively brief answers, although essentially correct. The guide to the length of answers is the lineage on the paper. The candidate clearly has the necessary knowledge and understanding to score well, but is unlikely to gain more than half marks.

Command words and phrases are the instructions that tell you what to do in exam questions. You must obey these command words. Let's take an example.

Understanding command words

> Describe the main features of a scree slope.

The command word 'describe' requires you to show knowledge of some of the main features of scree slopes. Your answer might refer to their steepness, the size and shape of scree particles, the location of scree slopes below rocky cliffs and so on. In a question of this type you must not *explain* how scree slopes develop. An examiner can only credit those parts of an answer that focus on *description*. Anything else will simply be ignored.

If, in our example, we replace 'describe' with the command word 'explain', the response should be quite different. Now you would need to give reasons for the angle of screes, the length of the scree slope and the distribution of particles by size on the slope. Your answer might refer to the height of the free-face, the size of particles and their angle of repose and the energy of falling rock particles and the frictional resistance to their downslope movement.

AS command words and phrases

Command word/phrase	Example	Instruction
State... Name ... Give ... Locate	Name one chemical weathering process which occurs on granite.	Write a short, clear answer (usually not more than one sentence).
Define ... What is meant by ...? Outline ...	What is meant by urbanisation? Define net migration change.	Write a brief answer that gives a succinct definition or meaning of a term or phrase. An example (a simple e.g.) will often help to clarify the answer.
Summarise ...	Summarise the global distribution of population change in the figure.	State the main trends, patterns, reasons and so on in relation to stimulus materials, issues etc. Most summaries will comprise a short paragraph.
Identify ...	Identify one factor that influences the steepness of beaches.	Pick out a relevant piece of information from stimulus materials you have been given.
Describe ...	Describe the spatial distribution of second homes in northwest England.	Give a word picture of the main characteristics of a distribution, feature, item etc. Your answer should not attempt to explain or account for the characteristics.
Compare ...	Compare the frequency distribution of slopes in region A (sandstone) and region B (clay).	Write about what is similar and what is different about two areas or two features.
Suggest reasons for ...	Suggest two possible reasons for the formation of podsol soils in area A.	Write down possible explanations. There may be other explanations that cannot be proved. Given the limits of the information provided, the reasons offered are likely to be tentative.
Explain ... Why ... ?	Explain why rates of weathering are most rapid in the tropical rainforest.	Give the causes or processes or reasons for a geographical phenomenon.
Examine ...	Examine the factors that influence the shape of the storm hydrograph in the figure.	Look critically at the main features (i.e. form, pattern, causes) of a geographical phenomenon.

Weathering and Slopes

Exam Question and Answer

Study the landform in the photograph and diagram. The area where the landform occurs is semi-desert. In July the average maximum daily temperature is 33°C; in December the average minimum temperature is –6°C. Mean annual precipitation is around 200 mm, much of it falling as thunderstorms in summer.

Shinarump formation
(resistant caprock)

DeChelly sandstone
(massively bedded,
pronounced vertical joints)   **B**

Organ rock shale   **A**
(relatively
unresistant)

a) Name the landform shown in the photograph.

Butte.

2/2

[2 marks]

b) Explain why slope A is less steep than slope B.

The shales forming slope A are softer than the sandstones forming slope B. As a result slope A is not as steep as slope B.

2/4

[4 marks]

c) Explain how rock structure might influence the retreat of slope B.

Slope B is made of sandstone and has vertical joints. The slope will retreat as large blocks of sandstone split along the vertical joints and topple onto slope A. Because the joints are vertical, slope B has a cliff-like profile.

0/4

[4 marks]

d) Name and explain one possible weathering process operating on slope B.

Freeze-thaw weathering will operate on slope B. This will occur in winter when average minimum temperatures are below zero.

3/4

[4 marks]

e) State and explain two possible mass movement processes operating on slope A.

1 Surface wash. Rain falling onto slope A will dislodge rock particles by rain splash. These particles will be moved downslope by water as it runs across the surface.

3/4

[4 marks]

2 Rock avalanche. Rocks weathered from slope B will accumulate on slope A. They may eventually become unstable and slide downslope as a rock avalanche.

4/4

[4 marks]

[Total 22 marks]

14/22

How to score full marks

Part a) Butte is correct (2 marks). Examiners would allow other possible features such as monadnocks *and* mesas (1 mark) **but not volcanic necks**.

Part b) The answer is in essence correct: you should expect 'softer' rocks to form less steep slopes than harder rocks. However, **the student fails to explain why this happens**. Without reference to the effects of weathering, erosion and mass movements on rocks the student fails to make the crucial causal connections and thus loses 2 of the possible 4 marks.

Part c) The answer deals with **lithology** (e.g. jointing) rather than **structure**. The points made by the student are valid, but are not appropriate to this question. Instead the answer should focus on the less resistant shale beds which, on erosion and weathering, undermine the overlying sandstones and cause them to topple along vertical joints. **The fundamental misunderstanding of the term 'rock structure' in this question means that no marks can be awarded for this answer.**

Part d) Two marks are awarded for correctly naming one possible weathering process. The introduction to the question makes it clear that frost, and presumably freeze-thaw weathering, occurs in this region. The student's explanation makes no reference to temperatures fluctuating above and below freezing (i.e. freeze-thaw cycles). **Simply stating that temperatures are below freezing is not sufficiently explicit and so the fourth mark is not awarded.**

Part e)

1 **Surface wash** is an important process on slopes in semi-arid environments (2 marks). The answer does not, however, explain the effectiveness of surface wash. There is no reference to the absence of vegetation or to the possible intensity of rainfall events (i.e. thunderstorms). The second part of the answer is a **description** rather than an **explanation**. **It is essential that the student responds precisely to the command word, in this case 'explain'.** No further marks can be credited for this answer.

2 Rock avalanches could occur in this area (2 marks awarded). The idea that slope A becomes unstable through rockfall and the accumulation of weathered rocks from slope B is also valid. The second part of the answer is just about worth 2 marks.

9

KEY FACTS

- **Weathering** (the breakdown of rocks exposed to fluctuations in temperature and moisture at or near the Earth's surface) and **erosion** (the wearing away of rocks and transport of rock particles by rivers, glaciers, wind and waves) are **two quite different processes**.

- **Rock structure** refers to (a) the juxtaposition of different rock types and (b) the angle of dip of rock strata.

- **Lithology** refers to the chemical composition, jointing, bedding and faulting of rocks.

- **Joints, bedding planes** and **faults** are lines of weakness in rocks which allow the ingress of water, air and plant roots. Weathering processes are concentrated along joints and densely jointed rocks are therefore particularly susceptible to weathering.

- When rates of weathering and rates of transport on a slope are exactly equal, the slope assumes a constant or **equilibrium** profile.

- **Slides** are the downhill transfer of regolith or rock moving as a uniform body across a slide plane. Velocity is constant throughout the sliding mass.

- **Flows** are the downslope movement of regolith where the velocity of the moving mass decreases with depth.

EXAM TECHNIQUE

- **Read the introduction to the question carefully: it may contain valuable information to help you answer the questions.**

- Make sure that you understand **what the question is asking you to do**.

- Take particular care with **geographical terms** such as 'structure' and 'lithology'. Inaccurate use of such terms will lose you many marks.

- The **depth** and **detail** of your answers must reflect the marks available for each question. As a rough guide you can assume there is one mark for each line of your answer.

Question to try

b) Define the term **mass movement**.

[2 marks]

b) State and explain two ways in which human activities can lead to mass movements and slope failure.

[4 marks]

c) Describe two methods which can be used to stabilise slopes and reduce the risks of mass movement.

[4 marks]

Study the diagram of landslides on Skye.

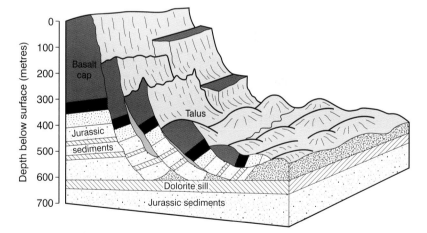

d) Identify the type of landslide shown in the diagram.

[2 marks]

e) Suggest one possible reason why landslides have occurred in this part of Skye.

[4 marks]

f) Explain how the geology shown in the diagram limits the base of the landslides.

[4 marks]

[Total 20 marks]

Answers are on page 81.

Exam Question and Answer

Figure 1 shows a storm hydrograph for the River Wyre in north Lancashire for a rainfall event in December 1993. The main features of the drainage basin are shown in Figure 2. The geology of the Wyre basin comprises impermeable shales and sandstones, and land use is dominated by permanent pasture, rough grazing and moorland.

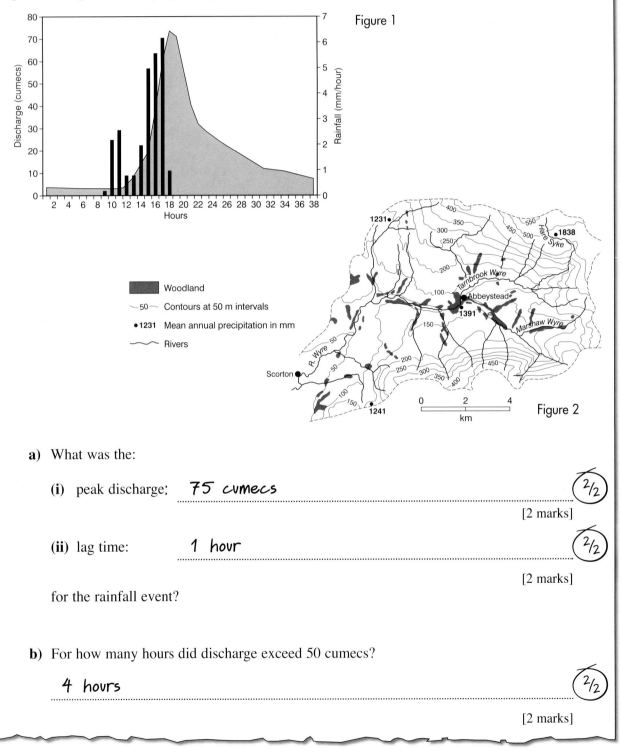

Figure 1

Figure 2

a) What was the:

   (i) peak discharge: **75 cumecs**

   [2 marks] $^2/_2$

   (ii) lag time: **1 hour**

   [2 marks] $^2/_2$

   for the rainfall event?

b) For how many hours did discharge exceed 50 cumecs?

   4 hours

   [2 marks] $^2/_2$

c) Why did the River Wyre not respond immediately to the start of rainfall?

Precipitation mainly falls on the land surrounding the river and it takes time for this to get to the river by throughflow and overland. (2/3)

[3 marks]

d) State and explain two reasons why, if the rainfall event of 19–20 December had occurred in June, the lag time would have been longer.

1 Vegetation would have had leaves, meaning that interception rates would be higher, slowing down the response time. (2/3)

[3 marks]

2 The ground would be drier in June so it would soak up more of the water before allowing it to travel to the river by throughflow. (1/3)

[3 marks]

e) Using only the evidence of Figure 2, suggest one possible reason why the River Wyre responds rapidly to rainfall events.

Its catchment area has little forest cover. This means that interception is low, precipitation reaches the ground quickly and water moves freely to the river. This movement is accelerated by the lack of tree roots which absorb moisture. (2/3)

[3 marks]

f) Explain why rivers such as the Wyre, with short lag times, often present a significant flood hazard.

When a river has a short lag time it is likely that peak discharge will be high. This is because the ground is not absorbing much water causing river levels to rise quickly and threaten flooding. The fact that the river responds quickly to precipitation means that humans have little warning of floods. The result is that damage is high as measures such as sandbagging cannot be effectively employed at such short notice. (2/6)

[6 marks]

[Total 24 marks] (15/24)

13

How to score full marks

Part a)

(i) – (ii) Both answers are accurate and score full marks. In questions of this type examiners normally allow some **latitude** in the **accuracy** of answers. In (i), 2 marks would be given for 75 ± 1 and 1 mark would be given for an answer in the range 75 ± 2. The lag time is exactly 1 hour (2 marks). An answer within a range of 30 minutes of 1 hour would score 1 mark.

Part b)

This answer (4 hours) is also correct and therefore worth 2 marks. One mark would be awarded for answers in the range of 4 hours ± 30 minutes.

Part c)

This answer is clear and accurate. However, the student could have provided a little **more detail** on how throughflow and overland flow delay the transfer of water to river channels. As a result the answer is worth 2 marks rather than 3.

Part d)

There is 1 mark for an accurate statement, with 2 marks for the explanation. The first answer is sound, but again the student **misses an opportunity to provide the necessary detail** on how interception increases lag time. The second answer is less satisfactory. The link between dry soil and the amount of throughflow is not fully explained. The student could have **developed the answer further** by referring to saturated soils and rapid overland flow in winter. Only 1 mark is credited.

Part e)

The student could have chosen slopes or drainage density, as well as forest cover. The last point (lack of tree roots, etc.) has little validity and has been ignored by the examiner. This means that **the relevant part of the answer is somewhat brief**. A fuller explanation is needed for 3 marks. Two marks are awarded.

Part f)

This question, worth 6 marks, is level marked (see the notes on page 5 of the Introduction for guidance on level-marked questions). The student's answer is neither detailed nor clear, and contains some misunderstanding. Although it has some validity, it lacks detail, accuracy and relevance and only achieves Level 1. The answer attempts to explain short lag-times, **which is not asked for**. The key idea – short lag-times cause rivers to exceed their bankfull capacities – has been overlooked. To gain the full 6 marks, the student **should have explained** that run-off is rapid and reaches the river channel quickly. The result is a very high but short-lived peak flow, which is likely to overtop the river's banks and cause flooding.

KEY FACTS

- **Hydrographs** describe the variable discharge of a stream or river in response to precipitation during a single storm, over a year, or over a number of years.

- The main features of a storm hydrograph are: the **lag time** (i.e. time delay between peak precipitation and peak discharge), the steepness of the **rising and falling limbs**, the area of storm flow, and the area of **base flow**.

- Flooding occurs when the discharge of a stream or river exceeds **bankfull capacity**. This happens during periods of prolonged precipitation (**slow flood**) or when precipitation is transferred into the river rapidly (**flash flood**).

- Drainage basins have both **fixed and variable characteristics** which influence the probability of flooding. Fixed characteristics include geology, slopes, vegetation cover, land use, soil types, and drainage density. Variable characteristics include the magnitude of precipitation events, the nature of precipitation (e.g. rain or snow), the amount of moisture in the soil, levels of evapotranspiration, seasonal variations in foliage and ground vegetation cover etc.

- Flood control comprises both **soft and hard engineering** responses. Soft engineering includes change of land use in the upper part of catchments and allowing rivers to flood across washlands which provide temporary storage. Hard engineering structures include flood embankments (levees), channel deepening, widening or straightening, dam construction etc.

EXAM TECHNIQUE

- Don't make **careless mistakes** when faced with simple tasks such as reading data from charts.

- To ensure accuracy in extracting data from charts you should be **prepared to use a ruler and if necessary draw guide lines on the chart**.

- You must **achieve a balance** by writing answers that are neither too brief nor excessively long. There should be just sufficient detail to gain maximum marks. Remember that the space left for your response is the best guide to the length of your answers.

- Make sure that your answers are always clear and **explicit**. Examiners can only reward what you have written: not what you could have written.

- You must always **apply your knowledge and understanding appropriately**. You must address the question set by the examiner, and not one that you would prefer to answer.

Question to try

The diagrams for this question are on the opposite page.

Figure 1 shows two bankfull channel cross-sections on the River Aire in North Yorkshire.

Figure 2 shows the upper catchment of the River Aire in the central Pennines, and the locations of the channel cross-sections. Most of the catchment comprises limestone, sandstone and shale, with a thick mantle of boulder clay. Mean annual precipitation in the Aire catchment is around 1200 mm.

a) Define the terms (i) bankfull discharge (ii) hydraulic radius (r).

[4 marks]

b) Explain why there is an increase in bankfull discharge between the upstream and downstream locations in Figure 1.

[3 marks]

c) Suggest one possible explanation for this difference.

[3 marks]

d) Explain why bankfull discharge, despite occurring on average on only one or two days a year, has a huge influence on channel shape.

[6 marks]

e) Describe and explain two flood control methods to reduce the risk of rivers exceeding bankfull discharge and comment on their disadvantages.

[10 marks]

[Total 26 marks]

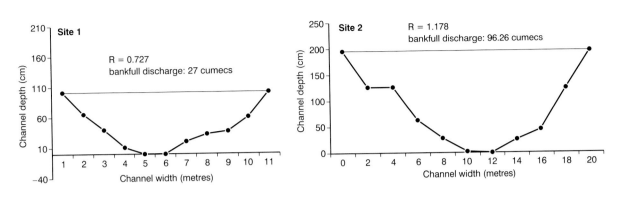

Figure 1

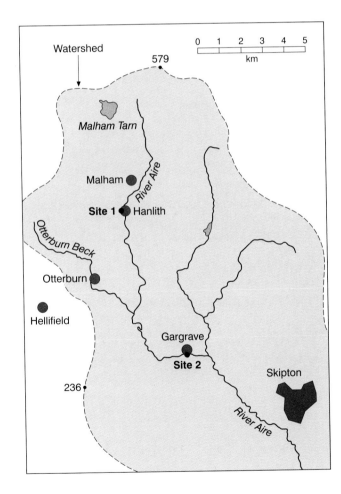

Figure 2

Answers are on page 82.

3 Coasts

Exam Question and Answer

Study the diagram which shows the cliffs at Blackhall in County Durham. The cliffs are between 40 and 50 metres high, and comprise horizontally bedded Magnesian Limestone, overlain by a thick layer of boulder clay. Both marine erosion and sub-aerial processes are active. As a result the cliffs have retreated by approximately 200 metres in the past 6000 years.

0/2

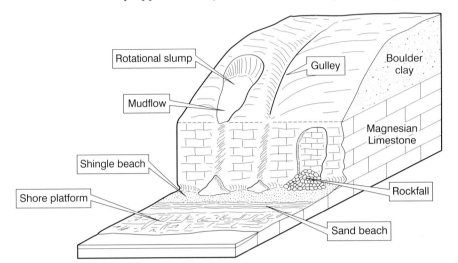

1/3

a) (i) Define the term **sub-aerial processes**.

> These are processes of weathering and mass movement. Weathering is the in situ breakdown of rocks by physical, chemical and biological processes. Mass movement is the downhill movement of materials as a coherent body under gravity.

4/4

[4 marks]

(ii) Identify two sub-aerial processes in the diagram.

> Rockfall and mudflow.

2/2

[2 marks]

b) Describe and explain two processes that are responsible for the retreat of the Magnesian Limestone cliffs in the diagram.

> 1 Undercutting by wave action. The hydraulic action of waves pounding the base of the cliffs at high tide loosens material, eroding the rock.

2/3

[3 marks]

2 Abrasion. The waves pick up shingle which break up the limestone by scouring the base of the cliffs.

(3/3)

[3 marks]

c) Suggest one possible reason why the Magnesian Limestone cliffs have a vertical profile.

Marine erosion cuts a wave-cut notch at the base of the cliff. The horizontally bedded rock above isn't strong enough to support itself, so it breaks off along a joint, leaving a vertical cliff.

(2/2)

[2 marks]

d) Explain why the boulder clay has a lower slope angle than the Magnesian Limestone.

The boulder clay slope angle is formed by sub-aerial processes such as slumping, slides, mudflows and gulleying. It is not undercut by wave action, so it has a lower angle.

(2/2)

[2 marks]

e) State and explain **two** possible reasons for the rotational slumping along the coastline at Blackhall.

1 The boulder clay has little coherence, so it is unable to sustain a steep slope angle. Slumping is a response to move towards equilibrium.

(2/3)

[3 marks]

2 As the cliffs retreat below the boulder clay loses its support and it slumps.

(2/3)

[3 marks]

f) Current shoreline management policy along the coastline at Blackhall is to do nothing to prevent erosion. Outline the possible advantages of this strategy.

> Local defence schemes such as groynes or sea walls only protect a small part of the coastline from erosion and may accelerate erosion in adjacent areas which are not protected.
> This is because stopping erosion may starve downdrift locations of sediment, and deplete their beaches. More erosion may increase inputs of sediments which build mudflats and decrease the probability of flooding along stretches of lowland coast.

3/6

[6 marks]

22/28

[Total 28 marks]

How to score full marks

Part a) This is a good answer. Two types of sub-aerial process are recognised and are **described accurately**. All 4 marks are awarded.

Part b) The first answer scores 2 marks out of a possible 3. The process of hydraulic action should be explained **in more detail** for full marks (e.g. the pressure exerted by water and air in rock joints). The second answer is very **succinct**, but its **accuracy** cannot be faulted and it just about achieves maximum marks.

Part c) The answer is worth the full 2 marks. Ideally there might be some reference to cliffs retreating parallel to the coastline.

Part d) The answer makes **an effective link** between processes and slope form and is therefore worth the full 2 marks.

Part e) This first answer is **not sufficiently explicit**. The **connection** between boulder clay's incoherence and the slumping process could be stronger. Two marks are awarded. The second answer is correct, but **very brief**. The answer requires a little **more development** to merit the third mark. There could be some reference to undercutting, cliff collapse and the resulting removal of the boulder clay's support. Again this is worth 2, rather than 3 marks.

Part f) This is a **level-marked** question (see the notes on page 5 of the Introduction for guidance on level-marked questions). The student describes two disadvantages. Both are relevant. However, the discussion requires **fuller development** to reach Level 3 (e.g. how the accretion of mudflats reduces the flood risk). **Detail**, applied correctly to the question, **is usually the quality which differentiates the top level**.

Don't forget:

KEY FACTS

- **Rock type** (lithology), **structure** (e.g. angle of dip of strata), **wave energy, sub-aerial processes** and **depth of water** offshore influence cliff profiles.

- Marine erosional processes include **abrasion, hydraulic action** and **corrosion**.

- Cliffed coastlines often show excellent examples of mass movement processes such as **rotational slides**, **mudslides** and **mudflows**. The undermining of cliffs by wave erosion is often the trigger to mass movement.

- **Cliff profiles** reflect a balance between geological factors, sub-aerial processes and wave energy.

- Weak, **incoherent rocks** such as boulder clay are most susceptible to sub-aerial processes. Marine erosion is relatively more important on **coherent rocks** such as limestone.

- **Coastal management** increasingly aims to work with natural processes rather than against them. It is a **sustainable policy** which allows erosion to occur in order to:
  (a) conserve the natural **sediment budget**
  (b) reduce the costs of building and maintaining **hard engineering structures**, such as sea walls and groynes.

EXAM TECHNIQUE

- In questions that require explanation, you **must emphasise** the cause-effect connections.

- Questions that are worth 6 marks are normally **level-marked** in three levels. Look at the notes in the Introduction, page 6, for more guidance on this.

- If your answer includes one or two sentences that show **clear understanding**, you will move to Level 2 (3–4 marks).

- The number of answer lines provided with a question gives a clear guide to the length of your answer. Answers that fill less than two-thirds of the space allowed are unlikely to have sufficient detail and will not score well.

- If your answer only shows quite **basic understanding** of what is being asked you will only achieve Level 1 (1–2 marks).

- If you include statements that show **detailed understanding** you will move to Level 3 (5–6 marks). Where appropriate, provide some **exemplification**. Often, this need not be more than a simple 'e.g. …'.

21

Question to try

Examiner's hints
- The command term 'what is meant by' and the command word 'define', are essentially the same.
- You must adhere strictly to instructions such as 'using only the evidence of Figure *a*'. The answer to such a question is in the stimulus material. No credit can be given for answers (however correct) which use prior knowledge.
- Questions worth 6 marks and above are level-marked. Bear in mind the qualities examiners look for in the highest level of response e.g. detail, accurate knowledge, appropriate use of knowledge and understanding. See the notes on page 5 of the Introduction for guidance on level-marked questions.

a) What is meant by the term **beach**?

[2 marks]

b) Add the following labels to the beach profile in Figure 1:
storm beach, berm, beach face, and breakpoint bar.

[4 marks]

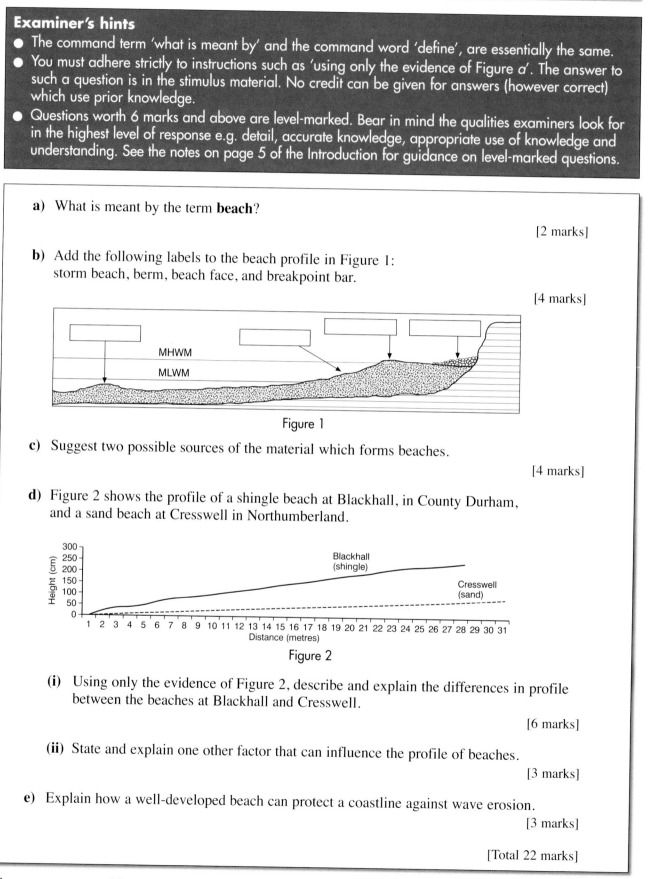

Figure 1

c) Suggest two possible sources of the material which forms beaches.

[4 marks]

d) Figure 2 shows the profile of a shingle beach at Blackhall, in County Durham, and a sand beach at Cresswell in Northumberland.

Figure 2

(i) Using only the evidence of Figure 2, describe and explain the differences in profile between the beaches at Blackhall and Cresswell.

[6 marks]

(ii) State and explain one other factor that can influence the profile of beaches.

[3 marks]

e) Explain how a well-developed beach can protect a coastline against wave erosion.

[3 marks]

[Total 22 marks]

Answers are on page 83.

4 Biogeography

Exam Question and Answer

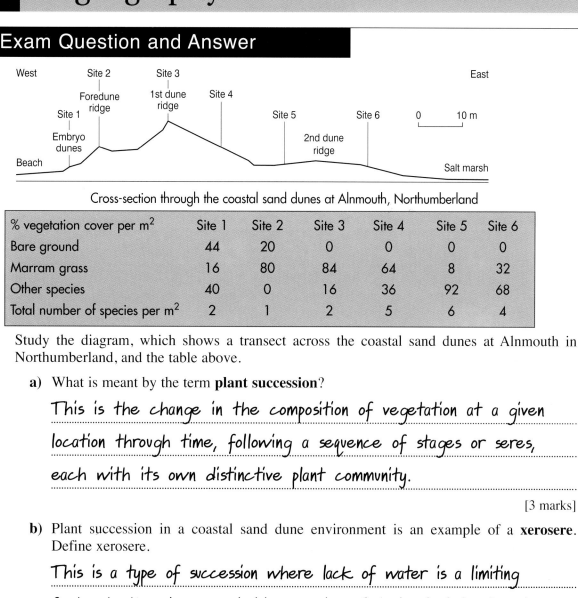

Cross-section through the coastal sand dunes at Alnmouth, Northumberland

% vegetation cover per m²	Site 1	Site 2	Site 3	Site 4	Site 5	Site 6
Bare ground	44	20	0	0	0	0
Marram grass	16	80	84	64	8	32
Other species	40	0	16	36	92	68
Total number of species per m²	2	1	2	5	6	4

Study the diagram, which shows a transect across the coastal sand dunes at Alnmouth in Northumberland, and the table above.

a) What is meant by the term **plant succession**?

This is the change in the composition of vegetation at a given location through time, following a sequence of stages or seres, each with its own distinctive plant community. (2/3)

[3 marks]

b) Plant succession in a coastal sand dune environment is an example of a **xerosere**. Define xerosere.

This is a type of succession where lack of water is a limiting factor to the pioneer colonising species. Examples include deserts and coastal dunes. (3/3)

[3 marks]

c) State and explain two possible reasons why nearly half of the area of the embryo dunes in site 1 on the diagram has no vegetation cover.

1 Rapid percolation and run-off may mean there is not enough water to sustain life. (1/3)

[3 marks]

2 High salt levels may create a hostile environment for plants to colonise, together with a lack of organic matter to provide nutrients. (1/3)

[3 marks]

23

d) Describe and explain two possible reasons why marram grass is the dominant species at sites 2 and 3 in the diagram.

1 Marram grass is hardy, withstanding wind and salt, so it is able to grow vertically in developed sand dunes to provide strength via its root structure.

(1/3)

[3 marks]

2 Marram grass is found at sites 2 and 3 rather than site 1, because it needs some organic matter provided by pioneer colonisers such as sand twitch.

(1/3)

[3 marks]

e) Explain why the number of different plant species shows a tendency to increase inland across the dune transect.

The further away from the sea the better the conditions because the dunes are more sheltered from the wind, are not affected by high tides, contain less salt and more organic matter. There are more plant species which thrive in these less harsh conditions compared to the embryo dunes. As a result competition develops with many different species colonising the older dunes.

(4/6)

[6 marks]

f) Describe and explain how human activities can cause accelerated erosion of coastal sand dunes.

Human activity may cause trampling, littering, grazing and fires in dunes and destroy the plant cover. This means that wind can erode the dunes more easily as there is no root structure. Gulleying and rilling also occur as there is less vegetation to intercept rainfall and reduce run-off.

(2/4)

[4 marks]

g) Outline one measure that can be taken to protect coastal dunes from such erosion.

Banning humans and livestock from dunes is an extreme measure which could be taken to protect the vegetation and reduce erosion. Even so, wildlife such as rabbits burrowing in the dunes would still have an erosional effect on the dune system.

1/4

[4 marks]

[Total 32 marks]

16/32

How to score full marks

Part a) The answer is accurate, but the student **misses an opportunity to provide detail** on the successional process e.g. increasing biodiversity and increasing biomass over time. The answer gains 2 marks rather than 3.

Part b) This is a **concise** and **accurate** answer which merits full marks.

Part c)
1 This answer **lacks development and detail**. Moreover, the student **refers unnecessarily** to two factors: percolation and run-off. The reference to percolation is worth 1 mark, but it is not supported by any explanation. Run-off does not occur on sand dunes. This inaccuracy is simply ignored.
2 Again **the student provides two answers, when only one is needed**. High salt levels and the absence of organic matter are both valid factors, but only one can be credited. The answers **fails to provide explanation**, and therefore the student loses 2 of the possible 3 marks.

Part d)
1 This is **a generalised answer** which is worth just 1 mark. For full marks the student should include such details as marram's extensive rooting system, its response to burial by wind-blown sand, its ability to conserve moisture etc.
2 This answer is **poorly structured**: the student suggests why marram is poorly represented at site 1 rather than focusing on its importance at sites 2 and 3. Examiners, however, mark positively and will ignore this mistake. The reference to organic matter is worth 1 mark, but its **lack of development** means no marks are awarded for explanation.

Part e) This question, worth 6 marks, is **level-marked** at three levels (see the notes on page 5 of the Introduction for guidance on level-marked questions). The answer is **clear** (rather than **detailed**) and therefore achieves Level 2 (4 marks). The amelioration of the environment with increasing distance from the shoreline is clearly stated. For Level 3, the answer should additionally give a **brief explanation** of the significance of shelter and organic material in the soil. The reference to high tides is factually inaccurate but would be ignored by the examiner.

Part f) Worth 4 marks, this question is also **level-marked** (two levels). The answer **does not have sufficient detail or accuracy** (e.g. gulleying and rilling are not found in sand dune environments) to achieve Level 2. To score full marks the answer would need to make explicit the activities which destroy vegetation (e.g. recreation and leisure) and make references to the reduction in wind speed and sand transport associated with a dense plant cover.

Part g) This question is **level-marked** (two levels). The answer is **generalised** and merely states that people and livestock should be excluded from coastal dunes. This is a Level 1 response which is worth 1 mark. **Specific** responses such as fencing to encourage sand accumulation and prevent further trampling, planting marram etc. should have been included, for the answer to gain full marks.

Don't forget:

KEY FACTS

- Coastal dunes form ridges parallel to the shoreline. The greater the distance from the shoreline the older the dunes.

- The older ('grey') dunes support soils with a higher organic and moisture content and provide a more sheltered environment for plants than the younger ('yellow') dunes.

- Changes in the environment with distance from the shoreline gives rise to a **zonation** of vegetation. The embryo dunes, foredunes and first dune ridge support few species and the vegetation cover, especially on the embryo dunes and foredunes, is sparse. The second and third dune ridges have a much greater **biodiversity, biomass**, ground cover and **productivity**.

- The vegetational changes that occur across sand dune environments are an example of **plant succession**.

- Coastal dunes are a **fragile environment**, easily damaged by human activity. Any destruction of the vegetation cover exposes the sand to wind erosion. Large areas of eroded dune are known as **blow-outs**.

- The main **limiting factor** for plant growth in coastal dunes is lack of fresh water. As a result plant succession in coastal dunes is a type of **xerosere** known as a **psammosere**. Other limiting factors include salinity (near the shore), lack of shelter, and the continual transport of sand.

EXAM TECHNIQUE

- Remember that a question worth 4 marks may be level-marked in two levels. To achieve Level 2 (3–4 marks), your answer must be **detailed and specific**.

- **In two-part questions which ask for description and explanation, you should answer each part separately**. This will ensure that you don't make **the most obvious mistake, which is to ignore explanation**.

- You must revise thoroughly so that your knowledge and understanding are accurate. Marks are not deducted for answers that are factually inaccurate. However, they are **self-penalising** because they take up valuable space (and time) and receive no credit.

- **Try to avoid writing generalised answers**. Generalisation usually implies a lack of detailed knowledge. Where appropriate, always include some specifics, even if they are no more than a simple 'e.g. ….'.

Question to try

Examiner's hints
- The command word 'name' requires no more than a simple one-word or one-phrase answer.
- Don't forget to use evidence from the diagram in your answer to question (d).
- In an extended-answer question (g), where a named example is required, it is not enough simply to name an example and then write generally about food chains or food webs. Detailed reference to an actual food chain or food web (e.g. temperate deciduous woodland, coral reef), individual species and transfers of energy are needed to achieve the highest level of response.

Study the diagram, which shows a desert food web in Death Valley, California.

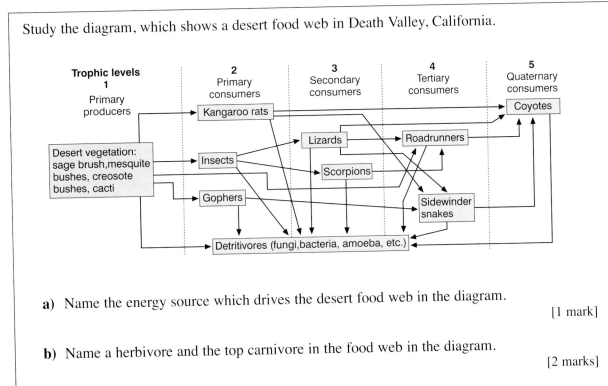

a) Name the energy source which drives the desert food web in the diagram.

[1 mark]

b) Name a herbivore and the top carnivore in the food web in the diagram.

[2 marks]

c) With reference to the diagram, explain why the roadrunner is classed as an omnivore.

[2 marks]

d) What is meant by the term **primary producer**?

[2 marks]

e) Explain why the biomass at each trophic level in the diagram will decline with distance from the site of primary production.

[4 marks]

f) Explain the importance of detritivores in the desert food web in the diagram.

[4 marks]

g) Describe and explain the main features of a food web (or food chain) in a named ecosystem you have studied.

[10 marks]

[Total 25 marks]

Answers are on pages 84–85.

Exam Question and Answer

Study Figure 1 which shows how atmospheric humidity varies with temperature.

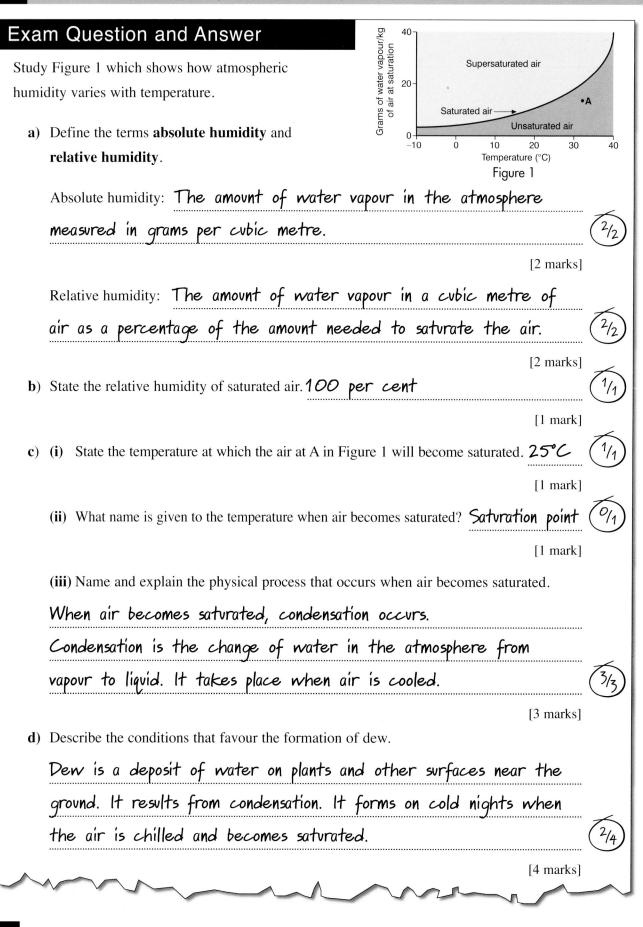

Figure 1

a) Define the terms **absolute humidity** and **relative humidity**.

Absolute humidity: The amount of water vapour in the atmosphere measured in grams per cubic metre.   (2/2)

[2 marks]

Relative humidity: The amount of water vapour in a cubic metre of air as a percentage of the amount needed to saturate the air.   (2/2)

[2 marks]

b) State the relative humidity of saturated air. 100 per cent   (1/1)

[1 mark]

c) (i) State the temperature at which the air at A in Figure 1 will become saturated. 25°C   (1/1)

[1 mark]

(ii) What name is given to the temperature when air becomes saturated? Saturation point   (0/1)

[1 mark]

(iii) Name and explain the physical process that occurs when air becomes saturated.

When air becomes saturated, condensation occurs.
Condensation is the change of water in the atmosphere from vapour to liquid. It takes place when air is cooled.   (3/3)

[3 marks]

d) Describe the conditions that favour the formation of dew.

Dew is a deposit of water on plants and other surfaces near the ground. It results from condensation. It forms on cold nights when the air is chilled and becomes saturated.   (2/4)

[4 marks]

e) Suggest one reason why dew is more likely to form in winter than in summer.

It is colder in winter than in summer so that condensation is more
likely to occur.

1/2

[2 marks]

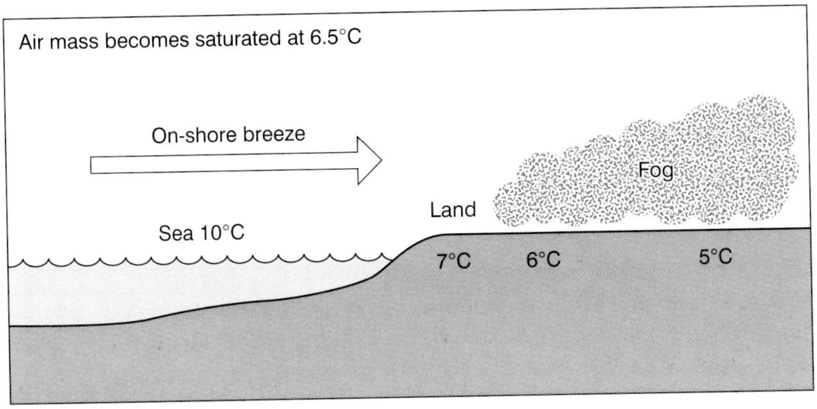

Air mass becomes saturated at 6.5°C

On-shore breeze

Sea 10°C

Land

7°C   6°C   5°C

Fog

Figure 2

f) Study Figure 2 which shows the formation of fog.

(i) Name the type of fog shown in Figure 2.

Advection fog

1/1

[1 mark]

(ii) Explain how the fog in Figure 2 formed.

Warm air from the sea moves onto the land. Because the land is
cooler than the sea, the air is chilled and condensation occur and
forms fog.

2/4

[4 marks]

g) Describe two ways in which fog disperses.

1 At sun rise, insolation warms the air and the fog evaporates.

2/2

[2 marks]

2 A warm air mass moves across the area. Temperatures rise and
the fog evaporates.

1/2

[2 marks]

18/25

[Total 25 marks]

How to score full marks

🎯 **Part a)** Both answers are **concise and accurate** and therefore score full mark[s].

🎯 **Part b)** This question requires a simple statement of knowledge. The answer is correct.

🎯 **Part c)**

(i) The student's answer demonstrates **understanding of the chart** and the **skill to extract accurate information** from it.

(ii) The answer simply repeats terms used in the question, so is unlikely to be what the examiner wants. No mark is awarded for 'saturation point'. The correct answer is 'dew-point'.

(iii) The answer correctly identifies condensation as the process. The **development** of the answer shows a **clear understanding** of how condensation takes place.

🎯 **Part d)** Marks are awarded for the definition of dew and the general observation that dew results from cooling and condensation. However, **a more detailed answer is required** for 4 marks. It should include a brief description of **specific** conditions favouring dew formation i.e. clear skies and heat loss, air that is already close to saturation before cooling occurs, light breeze etc.

🎯 **Part e)** This answer is not incorrect, but like (d) is rather **generalised**. For the second mark the student might have referred to longer nights in winter (i.e. more heat loss) and the higher relative humidity of air.

🎯 **Part f)** For a question attracting 4 marks, **the student should have provided more detail**. For example, the temperature information on Figure 2 has not been used. **Further development**, which would have secured full marks, could include references to the air temperature at sea, the decline of temperatures inland and the dew-point temperature of the air mass.

🎯 **Part g)** **The two processes cited are too similar to score full marks.** Both relate to a rise in temperature. A second factor, unrelated to temperature change (e.g. an increase in wind speed) would be appropriate.

Don't forget:

KEY FACTS

- **Humidity** is the amount of moisture in the atmosphere. **Absolute humidity** is the mass of water vapour in a unit volume of air. **Relative humidity** is the ratio between the actual amount of water vapour in the air and the amount needed to saturate it.

- **Condensation** is the phase change of water from vapour to liquid. It occurs when air is cooled to its **dew-point temperature**.

- **Air frost** is a temperature at or below zero above the ground. **Ground frost** is a temperature at or below zero at the ground surface.

- Condensation near the ground is responsible for **dew**, **fog** and **hoar frost**. The **cooling** that causes condensation is the result of:
  - either **radiation loss** from the Earth's surface at night
  - or the **horizontal movement of an air mass** across a cooler surface. High levels of relative humidity also favour condensation.

- **Evaporation**, caused by:
  - either a **rise in temperature**
  - or an **increase in wind speed**, disperses fog and dew.

EXAM TECHNIQUE

- Responses to the **command word** 'explain' must provide some **detail** about processes. Simply stating a process (e.g. condensation) is not enough to secure full marks.

- The marks allocated to a question give some idea of the **depth of detail** required. For example, for a question worth 4 or 5 marks, answers of just two or three lines are **inadequate**.

- When responding to questions that ask for two or three different reasons to explain a phenomenon or an event, there should be **no repetition or similarity** between the factors selected.

Question to try

Examiner's hints
- Take care not to confuse **depressions** with **anticyclones**.
- Remember that **definitions** do not require explanations.
- Study the weather chart and satellite image carefully, and match the two, **before** attempting the answer.
- **Areas of cloud** on a satellite image show **widespread cooling** of the atmosphere. This can only result from either rising air or advection.
- Latitude and longitude cannot be cited in answer to question (e). Your answer must be confined to the **evidence of weather** in Figure 1.

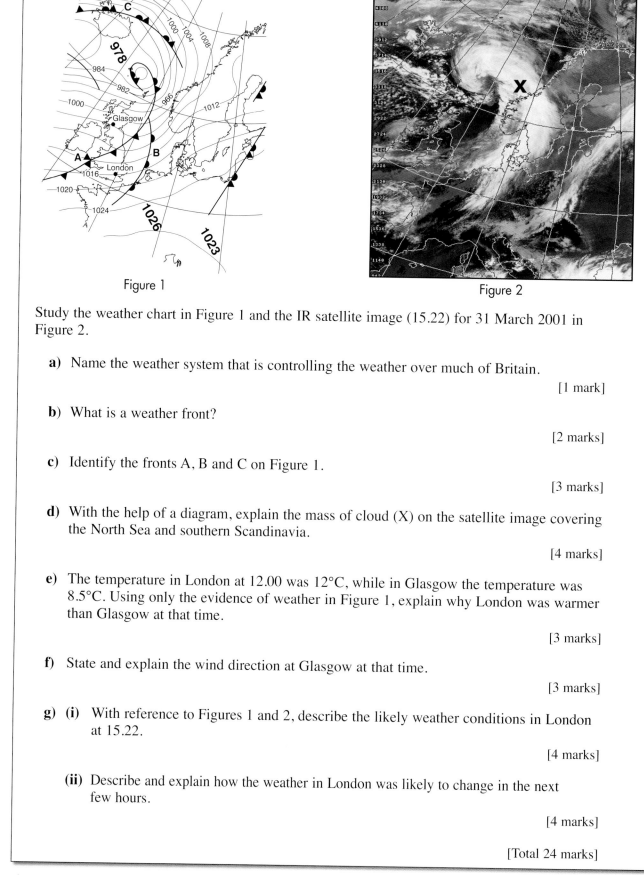

Figure 1

Figure 2

Study the weather chart in Figure 1 and the IR satellite image (15.22) for 31 March 2001 in Figure 2.

a) Name the weather system that is controlling the weather over much of Britain.

[1 mark]

b) What is a weather front?

[2 marks]

c) Identify the fronts A, B and C on Figure 1.

[3 marks]

d) With the help of a diagram, explain the mass of cloud (X) on the satellite image covering the North Sea and southern Scandinavia.

[4 marks]

e) The temperature in London at 12.00 was 12°C, while in Glasgow the temperature was 8.5°C. Using only the evidence of weather in Figure 1, explain why London was warmer than Glasgow at that time.

[3 marks]

f) State and explain the wind direction at Glasgow at that time.

[3 marks]

g) (i) With reference to Figures 1 and 2, describe the likely weather conditions in London at 15.22.

[4 marks]

(ii) Describe and explain how the weather in London was likely to change in the next few hours.

[4 marks]

[Total 24 marks]

Answers are on pages 85–86.

6 Plate Tectonics

Exam Question and Answer

Study Figure 1 and Figure 2 which show the tectonic plates and plate margins in the western USA and adjacent Pacific Ocean.

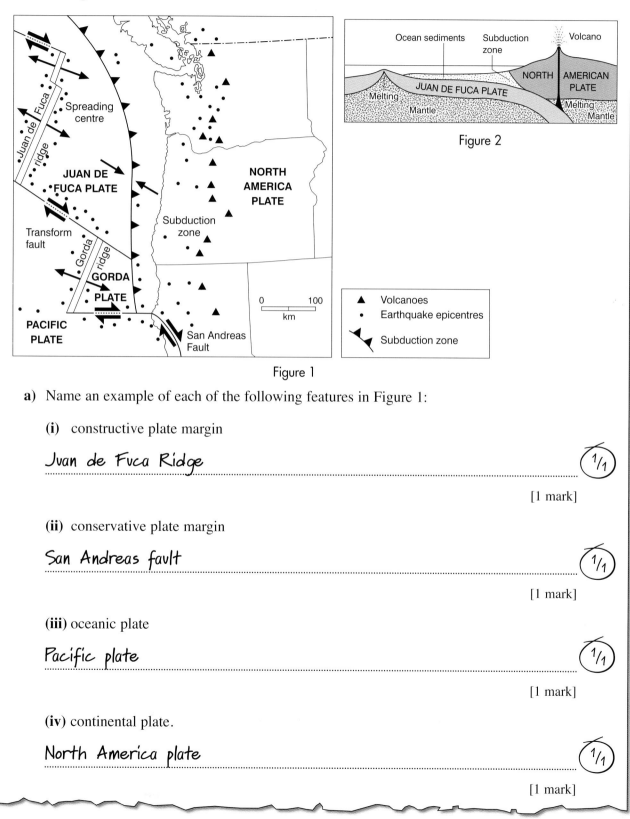

Figure 1

Figure 2

a) Name an example of each of the following features in Figure 1:

(i) constructive plate margin

Juan de Fuca Ridge   ①/₁

[1 mark]

(ii) conservative plate margin

San Andreas fault   ①/₁

[1 mark]

(iii) oceanic plate

Pacific plate   ①/₁

[1 mark]

(iv) continental plate.

North America plate   ①/₁

[1 mark]

b) Explain the south-eastward movement of the Juan de Fuca plate towards the North American plate.

The spread of the sea floor, caused by the movement of the Pacific plate, is pushing the Juan de Fuca plate from the north-west. This deflects the Juan de Fuca plate to the south-east. The Juan de Fuca plate is thus subducted because it is less dense than the North America plate.

1/4

[4 marks]

c) State and explain two pieces of evidence in Figure 1 that suggest that the Juan de Fuca plate is being subducted beneath the North American plate.

1 The decreasing size of the Juan de Fuca plate indicates that it is being subducted beneath the North America plate, where it melts.

0/3

[3 marks]

2 The volcanoes on the North America plate result from magma formed by the melting of the subducted Juan de Fuca plate. The magma forces its way to the surface to form a line of volcanoes.

2/3

[3 marks]

d) Give one reason for the subduction of the Juan de Fuca plate.

The Juan de Fuca comprises dense basalt. As a result it sinks below the lighter granitic rock which forms the North America plate.

2/2

[2 marks]

e) (i) Which plate boundary in Figure 1 is not associated with volcanism?

San Andreas fault

1/1

[1 mark]

(ii) Explain why there is an absence of volcanism along this plate boundary.

Because it is a conservative plate boundary, so the plates shear past each other, causing earthquakes but not volcanoes.

2/3

[3 marks]

12/20

[Total 20 marks]

How to score full marks

Part a) All of these answers are correct. As well as the Juan de Fuca Ridge, the Gorda Ridge is also a constructive plate margin.

Part b) The question is asking for an explanation of sea floor spreading. The answer refers to sea floor spreading (and therefore achieves Level 1, see the notes on page 5 of the Introduction for guidance on level-marked questions) but **provides no explanation of the processes involved**. The contention that the Pacific Plate is 'pushing' the Juan de Fuca plate is wholly inaccurate. For full marks there should be a **description** of the crustal tension along the Juan de Fuca Ridge, caused by magma rising from the mantle. New oceanic crust, formed along the Juan de Fuca Ridge, and convection currents in the mantle push the Juan de Fuca Plate south-eastwards towards the subduction zone.

Part c)

1 This part of the answer **does not use evidence from Figure 1**. Moreover, the statement that the Juan de Fuca plate is decreasing in size is incorrect. The student has overlooked the evidence of earthquake epicentres on the North America Plate. These earthquakes are caused by the irregular and jerky subduction of the Juan de Fuca Plate as it dives below the North America Plate and into the mantle.

2 This part of the answer **correctly cites the evidence** of volcanoes on the North America Plate and is therefore credited with 1 mark. A second mark is awarded for the link made between subduction and volcanism. There is **not quite sufficient detail** to justify 3 marks. Some reference to the movement of the **lighter** melt towards the surface would have secured the third mark.

Part d) This is a good answer: **precise and accurate**.

Part e)

(i) The San Andreas fault is **correctly identified** as a plate boundary without active volcanism.

(ii) The process of shearing is important but it **needs spelling out in more detail** for full marks. For example, shearing is a lateral rather than a vertical movement. It is the vertical movements (e.g. subduction, and rising convection currents along mid-ocean ridges) that generate magma and volcanism.

KEY FACTS

- Magma derived from the mantle rises to the surface at **mid-ocean ridges** or **constructive plate margins**.

- Tension in the crust causes rifting to occur at constructive plate boundaries. Following an **eruption**, **subsidence** leads to the formation of **rift valleys**.

- The evidence for **sea floor spreading** is the **alternating polarity** of rocks which form the ocean floor (i.e. iron particles in the basaltic rocks are aligned with the **Earth's magnetic field**).

- Sea floor spreading explains **continental drift**. The continents, surrounded by oceanic crust, move a few centimetres each year with the conveyor of sea floor spreading.

- **Subduction** occurs at **destructive plate margins**. Melt from the subducted (oceanic) plate rises towards the surface and creates **volcanic activity**.

- Subduction is not a smooth process. **Frictional drag** between the subducted plate and the surrounding mantle causes deep-seated **earthquakes**. Earthquakes also occur at **constructive plate margins** (the result of rifting) and **conservative plate margins**.

EXAM TECHNIQUE

- Accurate and detailed knowledge and understanding can only be achieved as the result of thorough preparation and revision. **Without a substantial knowledge base, you cannot hope to achieve the higher grades at AS level.**

- **Details of patterns and processes** often differentiate good answers from average answers. In extended-answer questions this detail may **include place specific examples** and named areas you have studied. In short-answer questions, the extra development which secures maximum marks may be no more than a sentence.

- When tackling short-answer questions it is essential to study the stimulus material carefully. **If a question asks you to use evidence from the stimulus material, you will receive no credit for answers (even if they are plausible) which simply draw on prior knowledge**.

Question to try

Study Figure 1, which shows the movement of the Indian sub-continent during the past 71 million years. This movement is known as **continental drift**.

a) Name the supercontinent in the southern hemisphere to which India was joined.

[1 mark]

b) How does the theory of plate tectonics provide an explanation for continental drift?

[6 marks]

c) Explain how the movement of continents may give rise to fold mountains.

[4 marks]

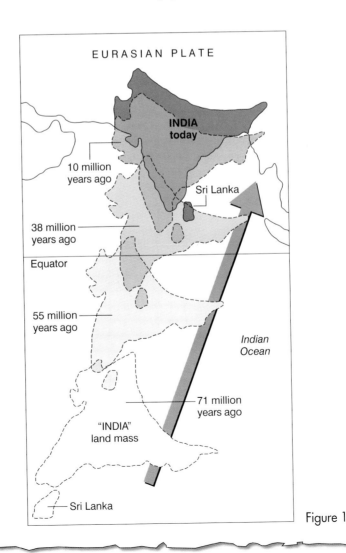

EURASIAN PLATE

INDIA today

10 million years ago

Sri Lanka

38 million years ago

Equator

55 million years ago

Indian Ocean

71 million years ago

"INDIA" land mass

Sri Lanka

Figure 1

Figure 2 shows a cross-section through a strato-volcano and types of volcanic hazards.

d) Name the features labelled A to E on Figure 2.

[5 marks]

e) With reference to a named example, explain the volcanic activity found at destructive plate boundaries. Use a diagram to support your answer.

[6 marks]

f) Outline the economic advantages of volcanoes.

[6 marks]

[Total 28 marks]

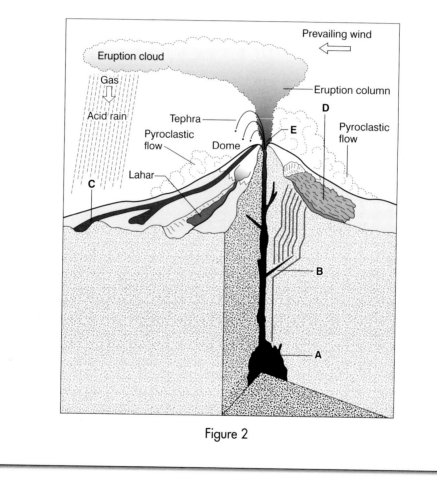

Figure 2

Answers are on pages 86–87.

Exam Question and Answer

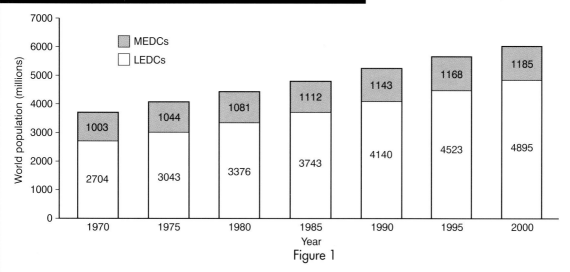

Figure 1

Study Figure 1 which shows changes in world population between 1970 and 2000.

a) Describe the main features of world population change between 1970 and 2000.

> There has been a deceleration in the growth of the populations of
> MEDCs from an increase of 41 million between 1970 and 1975,
> to 17 million between 1995 and 2000. LEDCs account for a
> much larger percentage of the world's population and their growth
> is accelerating from an increase of 339 million between 1970 and
> 1975, to an increase of 372 million in the period 1995 to 2000. ③/4

[4 marks]

b) Changes in the world's total population result from natural population change? What is meant by the term **natural population change**?   Natural increase

> This is the difference between the number of births in an area
> and the number of deaths. It excludes changes caused by migration. ②/2

[2 marks]

c) Name one factor, apart from natural population change, that might influence population change at a national or regional scale.

> People migrating to a region for work or a higher standard of living. ①/1

[1 mark]

d) State and explain the causes of population growth in LEDCs between 1950 and 2000.

Net migration into LEDCs is negative, so population growth must be caused by natural increase. Advances in basic medical care have increased in LEDCs because of work done by charities. This, together with better famine relief means life expectancy is rising. Meanwhile, contraception and family planning are still not commonly available, and mothers have large numbers of children who can contribute to family income at an early age. ⁴⁄₆

[6 marks]

e) Explain **either** for LEDCs **or** MEDCs, how population change can cause problems of dependency.

In LEDCs population increases caused by large families cause problems of dependency as farmland has to be divided into smaller plots according to the number of children. In subsistence regions such as NE India, this means that farmers cannot produce enough food to live on, and have to rely on international aid or state aid. ¹⁄₄

[4 marks]

f) Describe the population policies of a named country you have studied and comment on their effectiveness.

China's population increased from 560 million to 985 million between 1950 and 1980. The communist government believed that unless this growth was reduced, then supplies of food, water, energy and other resources would be threatened. As a dictatorship they introduced a 'two children per family' policy to halt the rapid growth of population. The penalties for exceeding the child quota were harsh economically, and abortion was encouraged.

Most people living in cities have accepted the policy, but in rural areas it is difficult to enforce, as labour is needed. Boys are also seen as more desirable, resulting in the killing of baby girls and unbalanced populations. Overall the proportion of young people has fallen, meaning there is less labour available. As the population ages there will be proportionally fewer workers to support the retired. As there is little state aid children have to support their aged parents. By 2030, 25 per cent of the population will be aged over 60, compared to just 9 per cent in 1980. Thus much more money will have to go to supporting retired people, causing a general fall in the standard of living.

(8/10)

[10 marks]

(19/27)

[Total 27 marks]

How to score full marks

- **Part a)** This is a Level 2 answer (see the notes on page 5 of the Introduction for guidance on level-marked questions). The student describes a number of valid changes and illustrates them with statistical evidence from Figure 1. Surprisingly, the most obvious change – the continuous and rapid growth of the world's population since 1970 – is overlooked.

- **Part b)** This answer is **accurate and precise** and scores the full 2 marks.

- **Part c)** The reference to migration is correct and is credited with 1 mark. The reason for migration is not required and is simply ignored.

- **Part d)** The answer correctly identifies natural increase as the cause of population growth in LEDCs between 1970 and 2000. It is also recognises the influence of both fertility and mortality in this process. However, the causes given for mortality decline are somewhat sketchy. A little **more knowledge** in this area (e.g. youthful age structures, better diets, application of modern medical technology etc.) would raise the answer from Level 2 to Level 3 (see the notes on page 5 of the Introduction for guidance on level-marked questions).

- **Part e)** This answer suggests that the student **lacks a sound understanding** of the concept of dependency as used in a demographic sense (i.e. children and elderly who are economically unproductive and depend on the adult, working population). The student achieves Level 1 and is credited for a reference to international aid which has some bearing on dependency.

- **Part f)** This is a detailed answer. For the most part it is accurate, and **handles both parts of the question** competently. In addition, references to **actual statistics and specific details** of China's population policies mean that this answer achieves the requirements for Level 3.

Don't forget:

KEY FACTS

- The **change in population** in a country or region results from the **numbers of births and deaths**, and the **level of migration**.

- **Rapid population growth** in **LEDCs** in the past 50 years is due largely to a **steep decline in mortality**.

- **Dependency** in a demographic sense means the **reliance of the non-productive part of a population** (i.e. children and old people) **on the productive population** (i.e. working adults).

- **Governmental population policies** aim to influence rates of population change. **Pro-natalist** policies encourage an increase in births and population growth; **anti-natalist** policies seek to reduce numbers of births and limit population growth. Some population policies are targeted at immigration and emigration.

EXAM TECHNIQUE

- Where appropriate, **always quote specific information (e.g. figures) from stimulus materials** such as charts, maps and diagrams.

- In extended-answer questions that require named examples of places or events, **reference to place specific details is essential. Generalised answers will achieve only modest scores**.

- The use of **accurate terminology** (e.g. fertility, mortality, age structure) is a feature of answers that achieve the highest marks.

Question to try

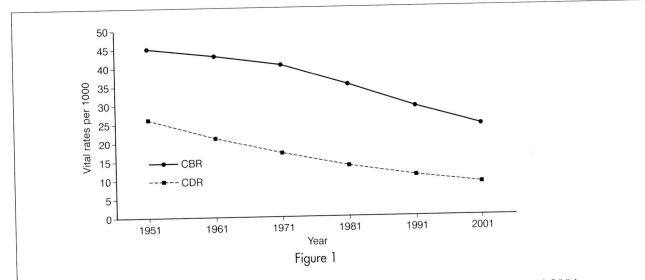

Figure 1

Figure 1 shows trends in crude birth rates and crude deaths in India between 1951 and 2001.

a) Define the term **crude birth rate**.

[2 marks]

b) Give one reason why the crude birth rate is not an accurate measure of fertility in a population.

[3 marks]

c) Using only the evidence of Figure 1, explain:

   (i) why India's total population increased almost threefold between 1951 and 2001

[2 marks]

   (ii) how India's population growth rate is likely to change in the next 20 or 30 years.

[2 marks]

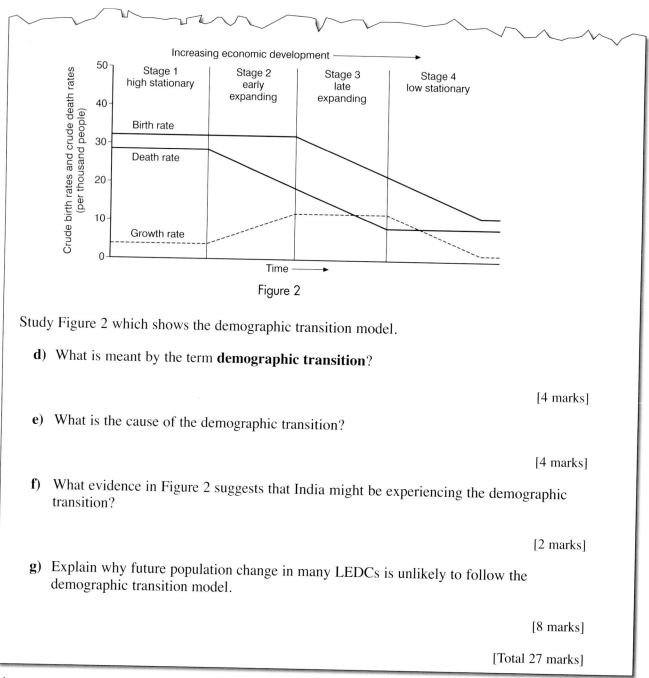

Figure 2

Study Figure 2 which shows the demographic transition model.

d) What is meant by the term **demographic transition**?

[4 marks]

e) What is the cause of the demographic transition?

[4 marks]

f) What evidence in Figure 2 suggests that India might be experiencing the demographic transition?

[2 marks]

g) Explain why future population change in many LEDCs is unlikely to follow the demographic transition model.

[8 marks]

[Total 27 marks]

Answers are on pages 88–89.

8 Migration

Exam Question and Answer

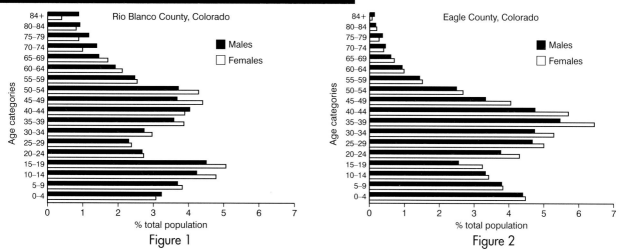

Figure 1

Figure 2

Figure 1 and Figure 2 show age-sex structures for Rio Blanco and Eagle counties in Colorado, USA. Migration has had a major impact on age structure in both of these rural counties.

a) What is meant by **migration**?

> Migration is the permanent or semi-permanent movement of people, changing their place of residence from one area to another. ②/2

[2 marks]

b) Describe and explain the evidence in Figure 1 which suggests that Rio Blanco county has suffered net migrational losses in the past 20 years or so.

> The decrease in population in the 50–54 year category upwards can be mainly attributed to mortality. However, the 25–34 age categories are substantially smaller than surrounding ones, indicating that these people are moving out. Because there are fewer women of childbearing age there are also fewer babies. The 15–19 age group is large as these people are less likely to have the means to migrate. ③/4

[4 marks]

c) What evidence suggests that migration in Eagle County has been age-selective?

> Without migration there should be a steady decrease in numbers in each age category, with increasing age. However, the age groups below 35 years are smaller than expected, indicating that young adults are migrating more than those in middle age. ⓪/2

[2 marks]

d) Why are people in some age groups more likely to migrate than others?

Young adults are most likely to migrate as they have no children at school, or elderly parents to tie them. They also often need to advance their careers. The elderly often have relatives and friends living close by so don't want to move. They are also less likely to have the resources to move. Children have no option to move. 18–20-year-olds often move away to university or college.

(3/6)

[6 marks]

e) With reference to a named example, explain how an unbalanced age structure can give rise to problems in rural communities.

Wensleydale in North Yorkshire is a good example of how migration has affected age-structure. Thornton Watless is a small village of around 50 houses, yet almost half the population are of pensionable age. The village is not close enough to a major town for commuting and, with few jobs available locally, the young have moved away. Meanwhile, the village is attractive for retirement, second homes and holiday homes. This has:

(a) added to the proportion of older adults in the population

(b) resulted in the closure of the post office and the recent reduction in the bus service to Bedale and Northallerton.

The community appears very close but, as the elderly die, more houses may be converted to holiday homes and second homes. This will put even more pressure on small businesses and services and cause hardship for those who are least mobile.

(6/8)

[8 marks]

(14/22)

[Total 22 marks]

How to score full marks

🎯 **Part a)** This definition correctly places the **emphasis** on the permanence of migration movements and is worth full marks.

🎯 **Part b)** It is reasonable to assume that the relatively small proportion of the population in the 25–34 age category is the result of net migrational loss. The student correctly suggests that this might explain the small proportion of children aged 0–4 years in the population. Although this is a Level 2 answer (see the notes on page 5 of the Introduction for guidance on level-marked questions), **some use of percentage figures from the charts is needed for maximum marks**.

🎯 **Part c)** This interpretation is **unlikely**. Rather than the small proportion of people aged below 35 years, the evidence for migration is the large proportion aged between 30 and 44 years. This suggests an age-selective in-migration, not an out-migration as the student assumes.

🎯 **Part d)** A Level 3 answer would include two or three **valid reasons** which explain how age affects mobility and migration. This answer correctly cites young adults as one group likely to migrate and old people as the group least likely to migrate. However, the answer provides only **limited development**. There are no references to economic factors such as employment, or social factors such as marriage. This is a Level 2 answer.

🎯 **Part e)** This answer is **detailed, accurate and place specific**. It successfully **shows the connection between demographic change and economic problems**, using the specific example of a village in the Yorkshire Dales. For full marks the answer needs further explanation of how second homes and holiday homes adversely affect rural services, and some precise information on, for example, the scale of population changein Thornton Watlass in the past 30 years or so.

Don't forget:

KEY FACTS

- Voluntary **migration is usually selective** by age, sex, income, employment etc.

- Migrations can be classified according to their **urban and rural origins and destinations** e.g. rural-urban migration, urban-rural migration etc.

- In **MEDCs** most migration is from **urban to rural areas**.

- **Young adults** are the age group **most likely to migrate** both in MEDCs and LEDCs.

- In **LEDCs** most migration is from **rural to urban areas**.

- In **MEDCs** migrant groups include **retirees and young families** moving from **urban to rural and semi-rural** areas. Both movements are examples of **counterurbanisation**.

EXAM TECHNIQUE

- **Explanations** based on **evidence** from charts, maps, diagrams etc. will only require a possible answer. Your answer may be inaccurate, but if it is **plausible**, given the data provided, it can score full marks.

- Questions which ask for explanations will probably require **at least two reasons in detail**, or **three or four reasons in less depth**, for maximum marks.

- It is essential to **plan your responses** to short-answer questions. With only a limited space for answers, you must be **precise, accurate and wholly to the point** in your responses.

Question to try

Examiner's hints
- Take care not to confuse **net migrational change** with **actual migration**.
- Remember that in descriptive questions such as (b) you should try to illustrate your answer with **information** taken from the **map** of migrational change in Britain.
- Make sure that you understand the distinction between the command words *describe* and *explain*.
- **Avoid mirror answers** to questions such as (d). If you described the impact on population growth of young people coming into a region, and their impact on the region they left, one answer could simply be the inverse of the other. In such a case you could only be credited with half marks.

Study Figure 1 which shows regional net migrational change in the UK in 1998.

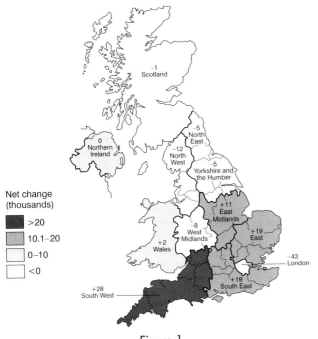

Figure 1

a) Explain what is meant by **net migrational change**.

[2 marks]

b) Describe the spatial pattern of net migrational change in the UK in 1998.

[4 marks]

c) State and explain the possible impact of age selective migration on future population growth in regions experiencing:

(i) net migrational gain

(ii) net migrational loss.

[6 marks]

Study Lee's model of migration in Figure 2.

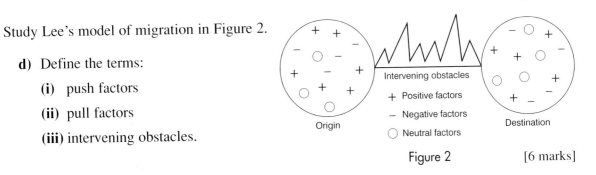

d) Define the terms:

(i) push factors

(ii) pull factors

(iii) intervening obstacles.

Figure 2

[6 marks]

e) With reference to migration within a named country or region, show how Lee's model can be used to understand migrational change.

[10 marks]

[Total 28 marks]

Answers are on pages 89–90.

Exam Question and Answer

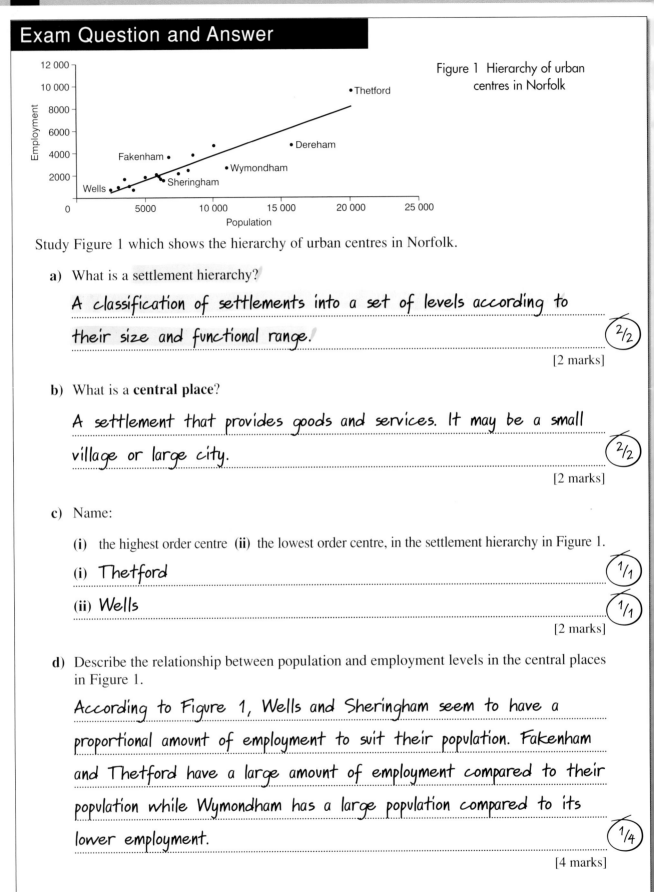

Figure 1 Hierarchy of urban centres in Norfolk

Study Figure 1 which shows the hierarchy of urban centres in Norfolk.

a) What is a settlement hierarchy?

> A classification of settlements into a set of levels according to their size and functional range.

[2 marks] ²/₂

b) What is a **central place**?

> A settlement that provides goods and services. It may be a small village or large city.

[2 marks] ²/₂

c) Name:

(i) the highest order centre (ii) the lowest order centre, in the settlement hierarchy in Figure 1.

(i) Thetford   ¹/₁

(ii) Wells   ¹/₁

[2 marks]

d) Describe the relationship between population and employment levels in the central places in Figure 1.

> According to Figure 1, Wells and Sheringham seem to have a proportional amount of employment to suit their population. Fakenham and Thetford have a large amount of employment compared to their population while Wymondham has a large population compared to its lower employment.

[4 marks] ¹/₄

e) Explain why higher order central places support more shops and a wider range of shops than lower order central places.

> Higher order places support larger numbers of shops as they need to accommodate their larger population and employment levels. They also need larger numbers of different shops for variety and to attract a range of people from other areas.

0/4

[4 marks]

f) Suggest one possible reason why Sheringham, a resort on the Norfolk coast, has less employment than average for its population size.

> I think that Sheringham has less employment than average for its population as it may have a large number of elderly people living there. Therefore only essential jobs will be needed. It may be a quiet village with many people living in it who do not require a wide range of amenities.

2/4

[4 marks]

g) Suggest one possible reason why the market town of Fakenham, has more employment than average for its population size.

> A reason why Fakenham may have more employment than average for its population size maybe due to the fact that people commute from nearby villages to sell their produce in the market. A very small number of people may live in Fakenham and yet travel to work there each day.

2/4

[4 marks]

11/22

[Total 22 marks]

Part a) This is a sound definition which is both **accurate** and **clear**.

Part b) The student might have added that central places serve not only their own populations, but also people who live in surrounding lower-order settlements. However, the answer demonstrates **sufficient understanding** for 2 marks.

Part d) A Level 2 answer will **recognise the overall positive relationship** in Figure 1, i.e. as population increases there is a proportional increase in employment in each urban centre. The relationship would then be demonstrated with examples taken from Figure 1. This answer focuses too much on individual urban centres and **loses sight of the overall pattern.** It therefore only achieves Level 1. (See the notes on page 5 in the Introduction for guidance on level-marked questions.)

Part e) The student needs to identify the **threshold concept** and apply it to the question, to achieve Level 2. The answer is **vague** and mistakenly implies that shops attract the necessary threshold population (it should, in fact, be the other way round).

Part f) The answer makes **only oblique reference** to Sheringham as a possible retirement centre and the subsequent limited demand for services. Although this is a valid reason, the connections between retirement and employment levels are **not made explicit**. Other possible answers include:

- Sheringham's coastal location which limits the size of its hinterland
- the seasonality of employment in many coastal towns.

Part g) This is a similar answer to that for part (f). It provides a valid reason but **fails to develop the connection** between employment levels and Fakenham's function as a market centre (accessibility, employment in service activities etc.).

Don't forget:

KEY FACTS

- **Settlement hierarchies** comprise large numbers of low-order centres (e.g. hamlets, villages) with progressively fewer centres at each higher level.

- **Central places** provide goods and services for their own populations and for people living in the surrounding **trade** (or **catchment**) **area**.

- On a scattergraph, a **best-fit trend line** which slopes from bottom left to top right indicates a positive relationship or **positive correlation** between the two variables.

- A best-fit trend line which slopes from top left to bottom right describes an **inverse** (or **negative**) **correlation**.

- **Threshold** is the number of people (or level of demand) required to support a function in a central place. Thresholds determine the total number of functions and range of functions found in central places.

EXAM TECHNIQUE

- If you have to describe a distribution or pattern (e.g. on a map or a chart) **start with broad generalisations**, which provide an overview. Then give the **detail**, quoting specific places, statistics and exceptions.

- If you are asked to give reasons to explain a trend, relationship or phenomenon, make sure that your suggestions are as dissimilar as possible.

- Your answers should always be **explicit**. Implying that there is a connection between two variables is not enough. It is your responsibility to **spell out answers clearly and unequivocally. Examiners will not fill the gaps for you**.

Table 1 Central places and changing functions in Lower Wharfedale: 1984–2001

Order	Central place	1	2	3	4	5	6	7	8	9	10	11	12
1	Arthington	✔✔	✔✗										
	Askwith	✔✔	✔✗										
	Clapgates	✔✗											
	Follifoot	✔✔	✔✔	✔✗									
	Goldsborough	✔✔	✔✗										
	Harewood	✔✔	✔✔	✔✔									
	Huby		✔✗	✔✔									
	Kirkby Overblow	✔✔	✔✗	✔✗									
	Kirk Deighton	✔✔											
	Leathley		✔✗										
	Little Ribston		✔✗										
	North Rigton	✔✔	✔✗										
	Sicklinghall	✔✔	✔✗	✔✗									
	Weeton		✔✗										
2	Burley	✔✔	✔✔	✔✗	✔✔	✔✗	✔✔						
	Collingham	✔✔	✔✔	✔✔		✔✗	✔✔						
	Pool	✔✔	✔✔	✔✔	✔✔								
	Spofforth	✔✔	✔✔	✔✔									
3	Otley	✔✔	✔✔	✔✔	✔✔	✔✗	✔✔		✔✗	✔✔	✔✔		
	Wetherby	✔✔	✔✔	✔✔	✔✔	✔✔	✔✔			✔✔		✔✔	
4	Harrogate	✔✔	✔✔	✔✔	✔✔	✔✔	✔✔	✔✔	✔✔	✔✔	✔✔	✔✔	✔✔

1 Pub, 2 PO, 3 Grocer/general store, 4 Chemist, 5 Independent shoes, 6 Independent clothes, 7 Multiple shoes, 8 Multiple clothes, 9 Boots, 10 Woolworths, 11 Gas/electricity showroom, 12 M&S

✔ = service present in 1984 and 2001   ✗ = service closed by 2001

Table 1 shows the hierarchy of central places in the Lower Wharfedale region of Yorkshire. This region forms part of the commuter belt for Leeds and Bradford.

a) State and explain one piece of evidence in Table 1 which suggests that a hierarchy of central places exists in Lower Wharfedale.

[3 marks]

b) Explain why most second-order central places in Table 1 support a larger range of functions than first-order central places.

[2 marks]

c) Outline the main changes in the provision of services in Lower Wharfedale (Table 1) which occurred between 1984 and 2001.

[4 marks]

d) State and explain two possible reasons for the changes in the provision of services in Lower Wharfedale (Table 1) between 1984 and 2001.

[6 marks]

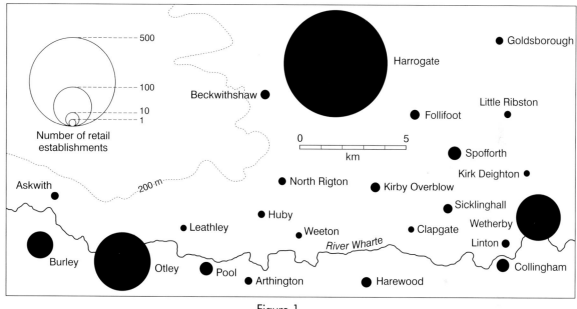

Figure 1

Study Figure 1.

e) (i) Describe the main features of the spatial pattern of central places in Lower Wharfedale.

[2 marks]

(ii) How does the concept of **range** help to explain the spacing of central places in Lower Wharfedale?

[4 marks]

f) With reference to a named rural settlement or rural region, describe recent changes that have occurred in service provision and explain how this has caused hardship for some members of the rural community.

[8 marks]

[Total 29 marks]

Answers are on pages 90–91.

Exam question and answer

Study Figure 1 which shows the spatial pattern of population change within the city of Austin, Texas, between 1980 and 1996.

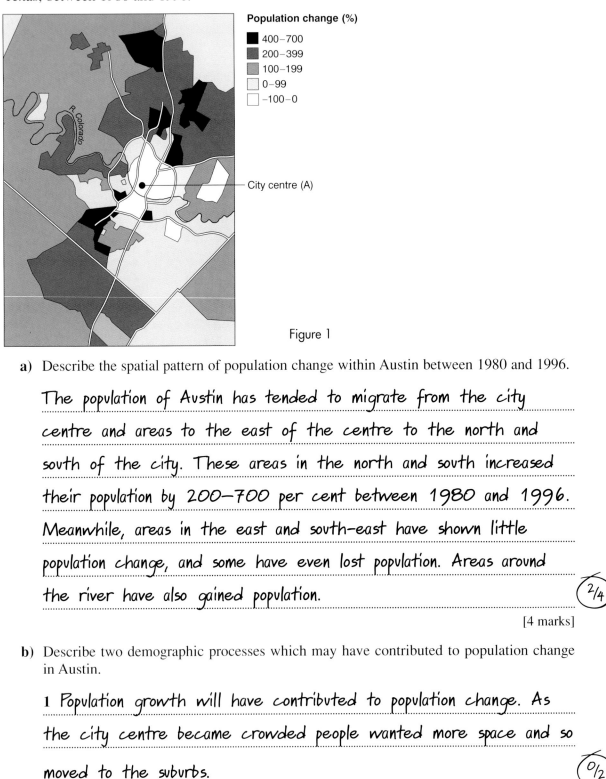

Figure 1

a) Describe the spatial pattern of population change within Austin between 1980 and 1996.

> The population of Austin has tended to migrate from the city centre and areas to the east of the centre to the north and south of the city. These areas in the north and south increased their population by 200–700 per cent between 1980 and 1996. Meanwhile, areas in the east and south-east have shown little population change, and some have even lost population. Areas around the river have also gained population.

2/4

[4 marks]

b) Describe two demographic processes which may have contributed to population change in Austin.

> 1 Population growth will have contributed to population change. As the city centre became crowded people wanted more space and so moved to the suburbs.

0/2

[2 marks]

2 Migration from other cities or countries will have caused
population change as there will be more and more people entering
the city and settling on the outskirts of the city.

2/2

[2 marks]

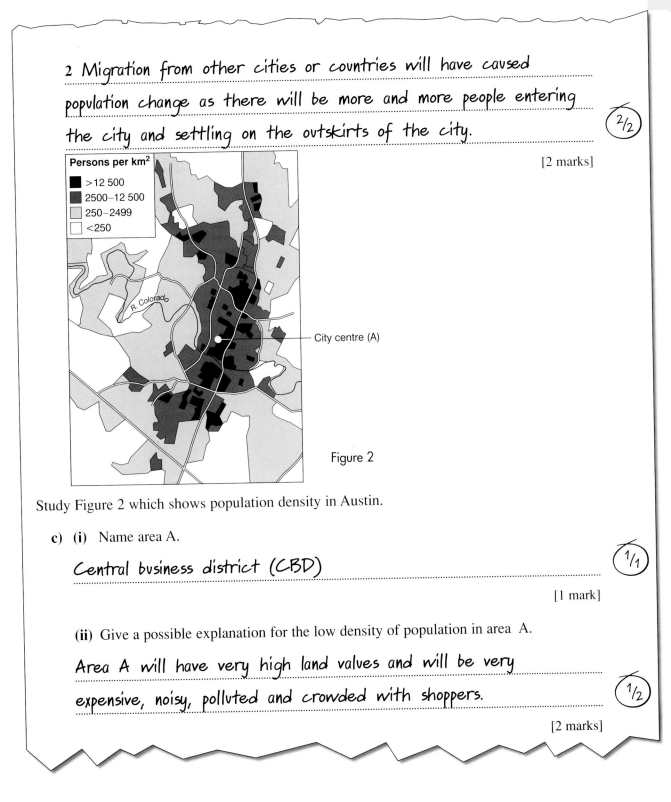

Persons per km²

- ■ >12 500
- ▨ 2500–12 500
- ▢ 250–2499
- □ <250

R. Colorado

City centre (A)

Figure 2

Study Figure 2 which shows population density in Austin.

c) (i) Name area A.

Central business district (CBD)

1/1

[1 mark]

(ii) Give a possible explanation for the low density of population in area A.

Area A will have very high land values and will be very
expensive, noisy, polluted and crowded with shoppers.

1/2

[2 marks]

d) Suggest two possible reasons why population density in Austin tends to decrease outwards from the central areas of the city.

1   People move out of the CBD to find a better quality of life. They set up in the suburbs where there is more space and better housing. As these areas become crowded they counter-urbanise and move further out to rural areas.   ⓪/2

[2 marks]

2   Population density decreases with distance from the CBD as there are fewer amenities and poorer transport networks. If people cannot travel and get to work they must live nearer the city centre.   ①/2

[2 marks]

e) Describe and explain the relationship between population density and population change in Austin.

The highest population density is in the CBD of Austin (up to 30000 per square mile). Yet Austin's CBD experienced a slight drop in population between 1980 and 1996. This is due to better transport systems. Even though people live in the CBD (which is densely populated) the better off can commute to the centre from the suburbs each day. The population changed dramatically between 1980 and 1996 in the north and south of Austin as people migrated to the suburbs either from the centre or from other cities or countries. Although population growth in these suburbs was as high as 700 per cent, the density of population in these areas is low. I think this is due to there being few people in the suburbs initially. Then with suburbanisation the change in population was rapid, though this has not yet led to high densities.   ⑤/8

[8 marks]

[Total 23 marks]   ⑫/23

58

How to score full marks

Part a) At the outset the student tries to explain population change (i.e. by referring to migration), even though the **command word** is 'describe'. The description which follows is valid, but **insufficiently detailed** to achieve Level 2 (see the notes on page 5 in the Introduction for guidance on level-marked questions).

Part b) Both answers invoke the **same demographic process** – migration. The student **does not consider a second** demographic process i.e. natural population change. Thus only 2 of the possible 4 marks are awarded.

Part c) **(ii)** The student loses a mark because there is no attempt to **explain** why high land values in the CBD cause low population densities.

Part d) The first answer does not **explain** the decrease in population density with distance from Austin's central area; it merely describes recent population movements. There is more **relevance** in the second reason, though the notion that most city dwellers work in the city centre is doubtful. Possible answers include the historical growth of cities, the type and density of housing, land values, family status and life styles etc.

Part e) This is a Level 2 answer (see the notes on page 5 in the Introduction for guidance on level-marked questions). It contains a number of **valid points**, identifying population decline in the city centre (where densities are high) and rapid growth in the outer suburbs (where densities are low). These are to some extent **illustrated** with **reference to specific areas** within Austin and rates of population growth. The answer, however, contains **little explanation**, and for this reason does not achieve Level 3. Confusion between the CBD and the inner city also means that the answer is **not entirely accurate**.

A Level 3 answer would have a **clear statement** of the spatial pattern of population change (decline or slow growth in the high density centre, rapid growth in the low density suburbs). This pattern would then be explained by migration movements. Intra-city movements would be influenced by push factors in the central areas and pull factors in the suburbs. In-migration to Austin could focus on the central area of the city, displacing longer established residents towards the suburbs. References to geographical areas within Austin and actual rates of population change would be expected in Level 3 answers.

Don't forget:

KEY FACTS

- The CBD is the central area of a city dominated by commercial land use and housing relatively few residents.

- The **inner city** is a zone of mixed land use, which includes high-density housing, industry and commercial activities, surrounding the CBD.

- **Suburbanisation** is a movement of people from the inner parts of cities to the outer suburbs. This is a movement from areas of high population density areas to areas of relatively low density.

- During the nineteenth century, in what are now MEDCs, most employment was concentrated in and around the city centre. Transport systems were poorly developed, forcing people to live at high densities close to the centre and near their places of work.

EXAM TECHNIQUE

- You must focus wholly on the **command words** of the question. It is a common mistake to 'describe' when asked to 'explain', and vice versa.

- **Don't waste time or space** in your answer by repeating the question. Short answers must be direct and to-the-point. Remember you are not writing an essay. However, you should **write coherently**, not in note form.

- If a question requires you to give **two reasons** or **two explanations**, they must be quite **separate and distinct**.

- Extended answer questions need careful structuring. Your ideas should be presented logically, the generalisations should be clear; and exemplification either from stimulus materials or from your own case studies should be used generously.

Question to try

Examiner's hints
- You should note that **family status** includes the concepts of family and life cycles.
- San Diego is a city in an **MEDC**, in California, the richest state in the USA: Tijuana is a city in Mexico, an **LEDC.**
- Possible explanations and reasons are often suggested by the **stimulus material**. The map of San Diego-Tijuana requires careful study.

Study the model in Figure 1, which shows the four factors which influence where people live in cities in MEDCs.

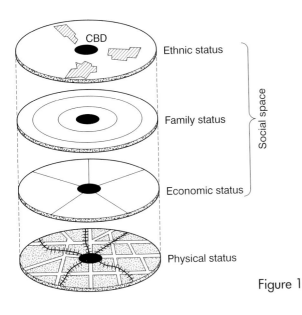

Figure 1

a) Define the terms (i) ethnic status (ii) family status.

[4 marks]

b) Explain how family status can influence where people live in cities in MEDCs.

[4 marks]

Figure 2 shows the distribution of the low income population in San Diego (USA) and Tijuana (Mexico). The two cities are adjacent to each other on opposite sides of the USA-Mexico border. Twenty-five per cent of San Diego's population is of Mexican origin. Of these, one-third migrated from Mexico to San Diego between 1990 and 1999.

Figure 2

c) Describe the distribution of low income groups in the San Diego-Tijuana region.

[4 marks]

d) Suggest **two** possible reasons for the concentration of low income groups within area A.

[4 marks]

e) Using only the evidence of Figure 1, state and explain one possible reason for the relatively small proportion of low income population in area B.

[3 marks]

[Total 19 marks]

Answers are on page 92.

Exam Question and Answer

Study Figure 1 which shows the employment structure of the USA, Greece and Pakistan.

a) Define the terms **primary** and **secondary** economic activities and give an example of each one.

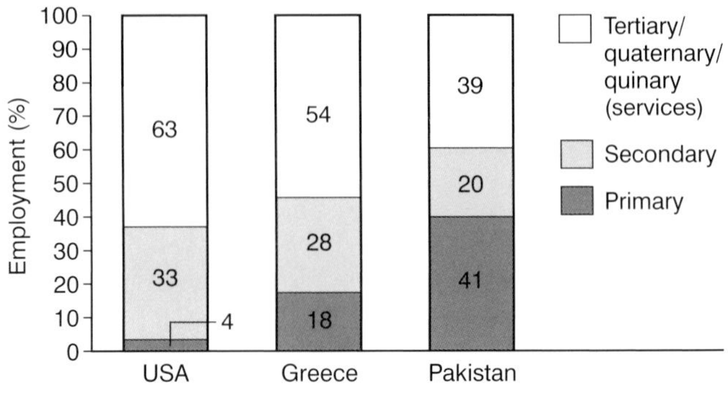

Figure 1

primary economic activity:

Produce food and raw materials. They are jobs which involve natural products.

2/2

[2 marks]

example: coal mining

1/1

[1 mark]

secondary economic activity:

the manufacturing sector including processing and assembly which add value to products.

2/2

[2 marks]

example: car industry

1/1

[1 mark]

b) Describe the main differences in the employment structures of the USA, Greece and Pakistan. (Refer to Figure 1.)

The USA has very low employment in the primary sector. Greece has a large proportion (around one-fifth) and Pakistan's employment in the primary sector is about two-fifths. The secondary sector in the USA and Greece is much larger than the primary sector. However, in Pakistan, the secondary sector accounts for only one-fifth of employment. The service sector employs over two-thirds

of working people in the USA. Greece has around half of its
employment in this sector, while Pakistan has nearly two-fifths in services. (2/4)

[4 marks]

c) Suggest one possible reason for the differences in the relative importance of service
activities in the three countries described in Figure 1.

Greece and the USA are more economically developed countries
(MEDCs) than Pakistan and therefore require more advanced
services such as those in the quaternary sector. Services such as
health care are more developed in Greece and the USA.
Because of greater wealth, MEDCs require more services. (2/2)

[2 marks]

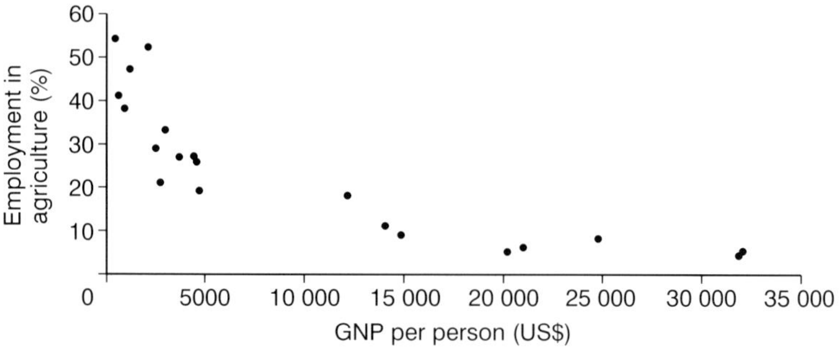

Figure 2

Figure 2 shows the relationship between male employment in agriculture and wealth (GNP per
person) in a sample of 21 countries.

d) (i) Describe the relationship between agricultural employment and GNP per person, as
shown in Figure 2.

Figure 2 shows that the wealthier a country, the less employment
there is in agriculture. This is because agriculture does not pay
high wages. (2/2)

[2 marks]

(ii) Give a possible explanation for the relationship between agricultural employment
and GNP per person in Figure 2.

Figure 2 shows that the relationship between agricultural
employment and GNP per capita is low. This could be due to

agriculture not being one of the more highly paid industries.

Competition from other economic sectors keeps wages low in agriculture. ⊘/4

[4 marks]

e) Describe and explain the principal changes in the employment structure, over the past 50 years, of a named country or region you have studied.

The UK's employment structure has gone through various changes in the past 50 years. The main employment in Britain used to be in the primary and secondary sectors e.g. coal mining, textiles etc. As coal resources ran out and industries such as textiles declined, new industries such as electronics developed. Meanwhile the service sector became the most important area of economic activity. Mechanisation caused a lot of unemployment in manufacturing industries, and many former factory workers moved to the service sector. Competition from foreign industries, which has been strengthened by improvements in transport and telecommunications, also hit Britain's manufacturing industries. 5/10

[10 marks]

[Total 28 marks] 17/28

How to score full marks

Part b) This answer is quite **detailed**, and **makes use of the statistics** in Figure 1. However, the answer tends to be little more than a list of features, **lacking an overview** or **summative qualities**. This approach suggests a Level 1 rather than aLevel 2 answer (see the notes on page 5 in the Introduction for guidance on level-marked questions).

Part c) Although the answer is repetitive, it states a **valid reason** and is therefore worth full marks.

Part d)

(i) The answer is **correct** and for this reason **gains both marks**. It is, however, **not perfect**. An ideal answer would refer to the inverse (or negative) relationship between the two variables, or state that as GNP per person increases there is a proportional decrease in employment in agriculture. The second sentence provides an **explanation**, albeit **irrelevant**. It is simply ignored by the examiner.

(ii) The first sentence is a description of the relationship. This is not required and, again, is ignored. The two sentences that actually **address the question** show little understanding of **GNP per person** as a general measure of the wealth of a country. The answer **fails to explain** how poor countries depend heavily on low productivity agriculture, while rich countries rely on more productive manufacturing and service activities.

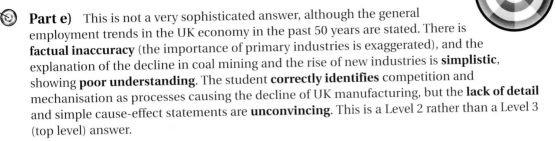

How to score full marks

🎯 **Part e)** This is not a very sophisticated answer, although the general employment trends in the UK economy in the past 50 years are stated. There is **factual inaccuracy** (the importance of primary industries is exaggerated), and the explanation of the decline in coal mining and the rise of new industries is **simplistic**, showing **poor understanding**. The student **correctly identifies** competition and mechanisation as processes causing the decline of UK manufacturing, but the **lack of detail** and simple cause-effect statements are **unconvincing**. This is a Level 2 rather than a Level 3 (top level) answer.

Don't forget:

KEY FACTS

- Because **service industries** in **MEDCs** are so diverse, and employ such a **large proportion of the workforce**, they are sub-divided into **tertiary**, **quaternary** and **quinary** sectors.

- The main difference in the **employment structures** of MEDCs and LEDCs is the size of the **primary sector**. While the primary sector in MEDCs is usually small, in many LEDCs it is the leading sector for employment and contributes significantly to GNP.

- **National employment** structures change over time with **industrialisation**, **rising wealth** and the eventual **creation of a post-industrial economy** dominated by **services**.

EXAM TECHNIQUE

- Although **misunderstandings, factual errors** and **irrelevancies** are not directly penalised by examiners, they often take up valuable space in short answers. Consequently, that part of an answer which is accurate and relevant may be insufficient to score high marks.

- **Extended-answer questions** provide an opportunity to write in some detail. To achieve the marks for the highest level of response, answers **must contain accurate detail** of places and processes as well as **sound general understanding**.

- Data plotted as scattergraphs are likely to show a **general trend**, but with some variation around a best-fit trend line. In addition to recognising the **overall trend**, you should **draw attention to the exceptions**. Remember that in geography, you rarely – if ever – encounter perfect positive and negative relationships between two variables.

Question to try

Examiner's hints

- Take care to **avoid 'mirror' answers** to questions which, for instance, ask for **two differences** between A and B. For example you can only score a maximum of half marks for stating that energy resource A is polluting, and energy resource B is non-polluting.
- **Primary energy production** relies on fuels such as **oil, coal, natural gas** etc. These fuels may be burned to generate electricity, which is known as **secondary energy**.
- You must respond to **comparison-type questions** with a **point-by-point** structure to your answer. You should never write two separate descriptions.

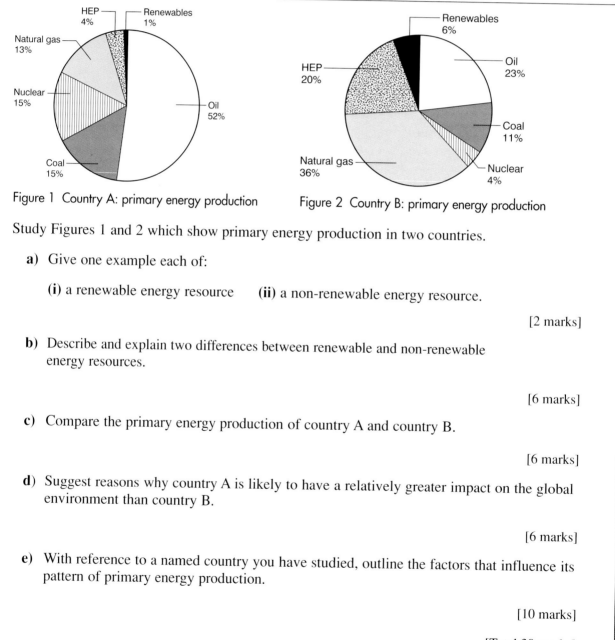

Figure 1   Country A: primary energy production

Figure 2   Country B: primary energy production

Study Figures 1 and 2 which show primary energy production in two countries.

a) Give one example each of:

    (i) a renewable energy resource   **(ii)** a non-renewable energy resource.

                                                              [2 marks]

b) Describe and explain two differences between renewable and non-renewable energy resources.

                                                              [6 marks]

c) Compare the primary energy production of country A and country B.

                                                              [6 marks]

d) Suggest reasons why country A is likely to have a relatively greater impact on the global environment than country B.

                                                              [6 marks]

e) With reference to a named country you have studied, outline the factors that influence its pattern of primary energy production.

                                                              [10 marks]

                                                              [Total 30 marks]

Answers are on pages 93–94.

Exam Question and Answer

Figure 1 shows a scree slope at Austwick in North Yorkshire. The screes have developed on Carboniferous Limestone, and the free face follows the line of the North Craven fault. Your task is to test the hypothesis that particles increase in size with distance down the scree slope.

Figure 1

a) Describe how you would:

(i) select lines of transect for the collection of data

Use a tape to measure the width of the scree slope along its base. This measured line is the baseline and runs horizontally across the foot of the scree slope. Use a calculator to generate five random numbers. Now take transects up the slope, at right angles to the baseline at five points selected at random. **4/4**

[4 marks]

(ii) select and measure scree particles along the lines of transect.

Divide each transect into five equal lengths. At the base of the transect and at the end of each length place a one metre square quadrat (with internal squares of 20 × 20 cm) on the scree. Select 25 scree particles from the intersections of the 20 × 20 cm squares. and measure their long, intermediate and short axes. **4/4**

[4 marks]

b) Name one type of chart you could use to represent the distribution of scree particles on the slope and justify your choice.

A scattergraph, with distance plotted on the x-axis and average particle size on the y-axis. This shows the distribution of particle size down the scree slope and allows conclusions to be drawn accurately. **1/3**

[3 marks]

c) Describe and explain one statistical technique you would use to analyse the the relationship between particle size and location on the scree slope.

You could use the standard deviation to show the average deviation from the mean particle size. The standard deviation would measure how closely the scatter of points on the chart followed a straight line trend. The lower the value of the standard deviation the better the fit of the trend line. This would measure the accuracy of the relationship between the size of scree particles and distance down slope.

4/10

[10 marks]

13/21

[Total 21 marks]

How to score full marks

Part a)

(i) This is a Level 2 answer (see the notes on page 5 in the Introduction for guidance on level-marked questions). It provides a **detailed and accurate description** of an objective and **feasible** method of sampling.

(ii) This answer also achieves Level 2. The method described is **practical** and is based on objective spatial sampling. Like the answer to (a)(i), it is probably **based on the student's own fieldwork**.

Part b) There is some **merit** in the scattergraph described in this answer, but it is **not the best way** to represent the data. A scattergraph presents particle size distribution as a single value for each sampling point on the scree slope. A **histogram** (or a frequency distribution curve) is a **better alternative** because it shows the distribution of particles in more detail.

Part c) For a question offering 10 marks, this response is **too brief**. In fact, the answer fills only half of the available space. The student suggests an analysis based on a regression (trend) line and the deviation of values from this line. This approach is **not unreasonable**, but the answer **lacks accuracy** and **fails** to show **detailed understanding**. The question is about correlation, which is the standard statistical measure of the relationship between two variables (i.e. particle size and distance downslope).

Don't forget:

KEY FACTS

- An **hypothesis** is a statement, the truth of which can be tested through scientific enquiry i.e. controlled data collection and objective analysis.

- The **objective** collection of data, through **scientific sampling**, is an essential step in geographical investigation.

- **Correlation** tests the **association** (or **relationship**) between **two variables** in a data set.

- The **U-test** and **t-test** examine the **differences** between two variables in a data set.

EXAM TECHNIQUE

- You should use **labelled sketch maps** and **diagrams** to **explain** methods of spatial sampling.

- The command word 'justify' indicates that you should **give reasons** for your answer.

- You should judge the required length of your answers by:
  - the **number of lines** or space allowed on the question paper, or
  - the **number of marks** allocated to the question.

Question to try

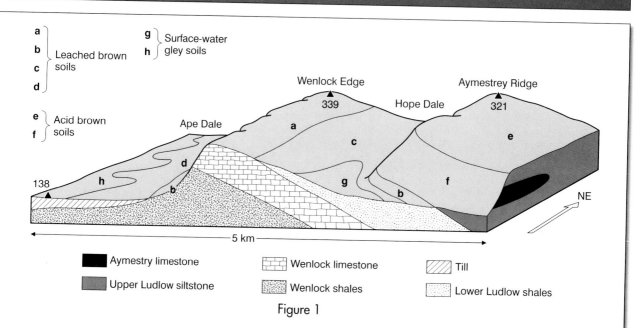

Figure 1

Study Figure 1 which shows rock types, relief and soils in the country around Wenlock Edge in Shropshire.

a) Using the information in Figure 1, state a possible question or hypothesis you could investigate through fieldwork.

[2 marks]

b) (i) Describe the sampling method you would use to collect data.

[4 marks]

(ii) Explain why you chose this sampling method.

[6 marks]

c) (i) Describe one type of graph you could use for representing the data you collect.

[4 marks]

(ii) State and justify one measure of central tendency you would use to summarise the data you have plotted graphically.

[6 marks]

[Total 22 marks]

Answers are on page 94.

13 Hypothesis Testing 2

Exam Question and Answer

Figure 1 shows the origins of shoppers, by place of origin, in the market town of Clitheroe in Lancashire. The data were collected by interviewing a random sample of shoppers in the town centre. The survey aimed to delimit the trade area of Clitheroe.

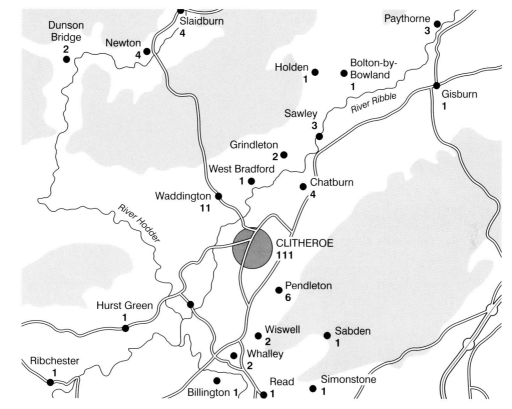

Figure 1 The origins of shoppers in Clitheroe

a) What is a **random sample** of shoppers?

This is when a sample of shoppers are interviewed, each shopper being selected randomly. This is achieved by sampling the nth shopper to pass the sampler, where n is chosen randomly by the random number generator on a calculator.

2/2

[2 marks]

b) (i) State and explain one advantage of random sampling.

A truly random sample gives a good representation of the population as every shopper has an equal chance of selection.

2/3

[3 marks]

(ii) Suggest an alternative method of sampling which could have been used in this study and outline its advantages over random sampling.

Systematic sampling could be used by selecting every nth shopper, where n is a fixed number. Systematic sampling is more practical and so easier than random sampling because it is unnecessary to generate a random number for every shopper.

3/3

[3 marks]

c) (i) Name one mapping technique which could be used to show the origins of shoppers in Figure 1.

Isoline map

0/1

[1 mark]

(ii) Give your reasons for choosing this mapping technique.

This mapping technique is easy to carry out and very effective. It gives a clear representation of the data from which good accurate results can be drawn.

0/1

[1 mark]

d) Explain how you would use the information in Figure 1 to test the hypothesis that the number of shoppers visiting Clitheroe declines with distance from the town.

Select a random transect starting at Clitheroe (i.e. by selecting a random direction between 1 and 360°) and draw the transect on the isoline map showing the origins of shoppers. Record the locations where the transect crosses each isoline, and the value of the isolines. A cross-section can then be drawn showing the relationship between the number of shoppers and distance from Clitheroe. The relationship between distance and the number of shoppers should be analysed using the Spearman rank correlation coefficient. The number of shoppers is likely to decrease with increasing distance from Clitheroe. This will give a negative correlation. The closer correlation coefficient is to −1, the stronger the relationship. A correlation coefficient close to zero suggests little or no relationship between the variables. If the correlation is high (i.e. close to −1) we can accept the hypothesis.

7/10

[10 marks]

14/20

[Total 20 marks]

How to score full marks

Part a) The answer **makes clear** the **essential point** that random sampling means selecting a sample using random numbers.

Part b)

(i) This answer needs a little **more development** to achieve full marks. For example, random numbers are selected objectively and therefore without bias.

(ii) Systematic sampling is often used as a simpler alternative to random sampling. The answer eventually **makes this point**, though rather indirectly.

Part c)

(i) Isolines are **not an appropriate mapping technique** to show the origins of shoppers. A flow map, showing numbers of shoppers by origin is more effective.

(ii) This answer is **highly generalised** and says nothing specific about the value of isoline mapping. Flow maps are designed to show the movement of people, freight and information. They show, in quantitative terms, the volume of and direction of flows.

Part d) **To achieve Level 3 the answer must explain accurately and in detail, a valid method** for testing the hypothesis that shopper numbers decline with distance from Clitheroe. The method proposed, though feasible, is unnecessarily complicated. Also because it depends on an isoline map and a single transect, accuracy may be compromised. A simpler method is to count the number of shoppers in distance zones around Clitheroe and correlate shopper numbers and distance. The strength of the answer is its focus on the Spearman rank correlation and its **correct application**. This just lifts the answer into the Level 3 category (see the notes on page 5 in the Introduction for guidance on level-marked questions).

KEY FACTS

- There are five statistical map types: **choropleth, dot, isoline, proportional symbols** and **flow maps**.

- In random sampling, every individual in a population has an equal chance of inclusion in a sample. **It does not mean selecting individuals without any conscious bias.**

- Scientific sampling is usually either **random** or **systematic**. If a population contains important sub-groups, they are more likely to be contained in a sample which is **stratified** i.e. a **random stratified** or **systematic stratified sample.**

EXAM TECHNIQUE

- Where a short-answer question contains **two command words** such as 'state' and 'explain' it is simplest and clearest to **organise your answer into two separate parts.**

- If you are asked to 'give reasons', **two or three, investigated in some detail, should ensure full marks.** Any attempt to deal with **more than three will probably result in superficial coverage and some loss of marks.**

- You should avoid giving answers that take up more than the allowed space. Your answers should **just fill the lines on the question paper.**

Question to try

Examiner's hints
- Your **hypothesis** and **subsequent investigation** must be at the **scale** of Figure 1.
- Keep your hypothesis as **simple** as possible. Ideally it should involve **no more than one or two variables**.
- The command word 'outline' asks you to **describe something fairly briefly**.
- This question **does not ask for details** of questionnaire surveys (format, postal, doorstep etc.), traffic counts or other methods of data collection.

Figure 1

Figure 1 is based on a 1:10 000 map which shows part of a suburb on the edge of a British town. Your task is to investigate the commuting behaviour of residents in this suburb.

a) State and justify one hypothesis you might test in an investigation of commuting from the suburb in Figure 1.

[3 marks]

b) Assume that your investigation is based on a sample of 100 households in the suburb.

(i) Why is it important to collect data from a large sample of households?

[2 marks]

(ii) Describe how you would select a systematic sample of households from the suburb in Figure 1 to test the hypothesis you have chosen.

[4 marks]

(iii) Outline the possible strengths and weaknesses of systematic sampling in this investigation.

[6 marks]

c) Describe an alternative method of sampling you could use to test your chosen hypothesis and explain how it would assist in your investigation.

[6 marks]

[Total 21 marks]

Answers are on page 95.

Exam Question and Answer

Table 1 shows data based on a sample of 100 dolerite erratic blocks found on the shore platform at Blackhall on the coast of County Durham.

0.065	0.103	0.070	0.050	0.072	0.254	0.058	0.144	0.229	0.082
0.055	0.054	0.029	0.039	0.050	0.061	0.098	0.107	0.070	0.065
0.037	0.156	0.039	0.050	0.086	0.032	0.085	0.032	0.078	0.044
0.113	0.039	0.057	0.042	0.094	0.216	0.069	0.046	0.434	0.056
0.027	0.084	0.092	0.022	0.196	0.056	0.041	0.068	0.064	0.036
0.039	0.067	0.044	0.116	0.063	0.070	0.179	0.055	0.097	0.060
0.063	0.086	0.034	0.051	0.213	0.168	0.041	0.219	0.072	0.152
0.143	0.061	0.085	0.042	0.092	0.072	0.084	0.022	0.060	0.082
0.025	0.112	0.078	0.046	0.088	0.061	0.028	0.038	0.050	0.057
0.147	0.115	0.039	0.097	0.409	0.041	0.047	0.039	0.050	0.039

Table 1 Size (m^3) of dolerite erratics at Blackhall

a)   Complete the frequency table (Table 2) to show the distribution of dolerite erratics.

[2 marks]

2/2

Classes (m^3)	Frequency	
< 0.05	29	
0.05 – 0.069	?	*(27)*
0.07 – 0.089	17	
0.09 – 0.109	?	*(7)*
0.11 – 0.129	4	
0.13 – 0.149	3	
0.15 – 0.169	3	
> 0.169	9	

Table 2   Frequency table for dolerite erratics

b)   Plot the data from the frequency table on the chart below.

Frequency distribution of dolerite erratics at Blackhall

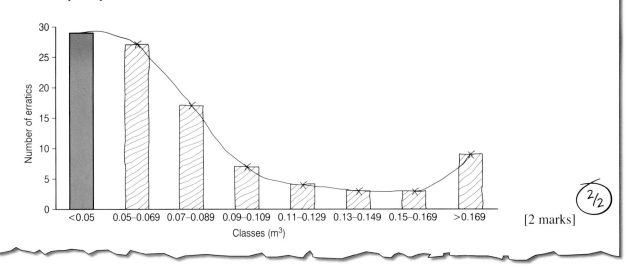

[2 marks]

2/2

c) Name the type of chart you have drawn. Histogram

[1 mark]

(1/1)

d) Plot a frequency distribution curve on Figure 1.

[2 marks]

(2/2)

e) Describe the frequency distribution of erratics in the chart.

Most erratics are small with the largest number (i.e. 27 per cent) being less than 0.05 m³. There is a gradual decrease in frequency as size increases, to a low point in the 0.13 to 0.169 m³ categories. Frequency then increases to 9 per cent of erratics over 0.169 m³.

[3 marks]

(2/3)

f) (i) Name and justify one measure of central tendency which could be used to describe the data in the chart.

The median could be used. This is because the frequency distribution is positively skewed and so a measure of central tendency that gives equal emphasis to each value is more accurate than any other measure (e.g. the mean, which takes account of the magnitude of each value).

[4 marks]

(4/4)

(ii) State and explain two disadvantages of your selected measure of central tendency for describing the data in the chart.

One disadvantage is that when using the median, frequency distributions that are drastically different can have the same median value. So two distributions, one positively skewed and one negatively skewed could have the same medians. Distributions cannot therefore be compared accurately.

[3 marks]

(3/3)

g) Describe one measure of dispersion that could be used to summarise the spread of values in Figure 1 and justify your choice.

One measure of dispersion is the inter-quartile range. This uses the median which has already been calculated. If the median splits the distribution in half (50 per cent either side) then the median of each half of the distribution splits the distribution into quarters (or quartiles). The values which separate these quarters are known as the upper and lower quartiles. The inter-quartile range is the difference between the upper and lower quartiles. The inter-quartile range uses the median which we have seen is the best measure of central tendency for the data in the chart.

4/5

[5 marks]

20/22

[Total 22 marks]

How to score full marks

Part e)   This answer could be improved with **a description of the overall shape of the frequency distribution**, and some reference to its **asymmetry** or **skewness**. The use of figures taken directly from the chart is credited, though the answer becomes something of a listing of features.

Part f)

(i)   This answer clearly reaches Level 2 (see the notes on page 5 of the Introduction for guidance on level-marked questions). It names an appropriate measure of central tendency for 1 mark, and then provides a detailed justification. Reference to the frequency distribution's positive skew and its implications, is impressive.

(ii)   To achieve full marks an answer must state two disadvantages and deal with at least one in some detail. **Although this answer is detailed, it considers only one disadvantage.** It is, therefore, worth 2 marks rather than 3.

Part g)   Knowledge and understanding of inter-quartile range are competent and gain a full 3 marks. But the justification for its use focuses on **a single point and comprises just two lines of the space allowed for the answer**. This requires a little more development (or the introduction of a second point of justification) for full marks.

Don't forget:

KEY FACTS

- Frequency distributions may be **normal** (i.e. symmetrical) or **skewed** (i.e. asymmetrical).

- **Frequency distributions** may be represented graphically by **histograms** and **frequency curves**.

- Measures of central tendency include the **mean**, **median** and **mode**. Choice of measures of central tendency depends on the nature of the frequency distribution.

- Measures of dispersion are used to show the spread of data around a central value (e.g. mean, median). They include the **range, inter-quartile range** and **standard deviation**.

EXAM TECHNIQUE

- Extended-answer questions require some planning. They should be structured (like a mini-essay) and should be written in **continuous prose**.

- The use of **bullet points** in both short-answer and extended-answer questions is permitted, but **responses must be coherent and not just lists of points**.

- Although questions on the techniques of geographical investigation are often very specific, **often more than one technique is appropriate**. This applies to mapping methods, statistics, charts and sampling procedures.

Question to try

Examiner's hints
- You should relate the strengths and weaknesses of statistical maps to the specific task of mapping the population of north Norfolk.
- Remember that the figures in Table 1 are **population totals** and **not population densities**.
- Think about the suitability of different statistical maps for showing:
  - point data or area data
  - continuous or discrete spatial distributions.

Table 1 Population of selected parishes in north Norfolk

Baconsthorpe	182	Hempstead	165	Salthouse	198
Beeston Regis	1087	High Kelling	515	Sheringham	5787
Blakeney	870	Hindolveston	412	Stody	169
Bodham	361	Holt	3380	Swanton Novers	235
Briningham	128	Itteringham	157	Thornage	207
Briston	1602	Kelling	191	Thurning	51
Cley	517	Langham	347	Upper Sheringham	260
Corpustry	626	Letheringsett	255	West Beckham	205
East Beckham	33	Little Barningham	110	Weybourne	557
Edgefield	361	Matlask	140	Wickmere	114
Field Dalling	287	Melton Constable	532	Wiveton	152
Gresham	414	Morston	532		
Gunthorpe	231	Plumstead	145		

The data in Table 1 form part of an investigation into the population geography of north Norfolk. You are required to use these data, and the map in Figure 1, to **explain how to** construct a dot map of the distribution of population in north Norfolk.

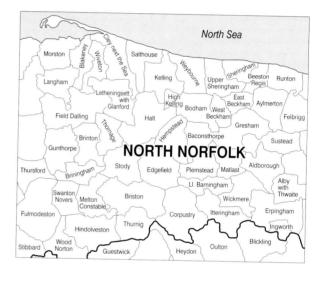

Figure 1 North Norfolk

a) What is a dot map?

[2 marks]

b) When constructing a dot map, explain how you would decide:
   (i) the value of each dot   [4 marks]
   (ii) the placement or positions of dots on the map   [4 marks]

c) State and explain one advantage of using a dot map to represent the data in Table 1.

[3 marks]

d) State and explain two disadvantages of using a choropleth map to represent the data in Table 1.

[6 marks]

e) Name and justify one other mapping technique which might be used to represent the data in Table 1.

[3 marks]

[Total 22 marks]

Answers are on page 96.

1 Weathering and Slopes

How to score full marks

a) Mass movement is the downhill transfer of slope materials moving as a coherent body. These movements, caused by the effect of gravity on slopes, comprise slides (e.g. mudslide), flows (e.g. solifluction) and heaves (e.g. soil creep).

b) 1 Deforestation of slopes. The removal of forest cover on slopes reduces slope stability. Tree roots bind the regolith and resist the downslope force of gravity. Trees also intercept rainfall and, through transpiration, reduce soil moisture. High moisture content is a common cause of slope failure.

 2 Loading of slopes. Buildings may be constructed on slopes, or waste materials may be dumped on slopes. The effect of loading is to increase the mass of the slope and the gravity force. Slope failure and mass movement may result.

c) 1 Terracing. Benches can be cut on slopes to allow building or cultivation. Terraces will slow the movement of water down slopes and reduce the effect of the gravity force.

 2 Reduction in slope angles. The steepness of slope profiles can be reduced. These slopes then support angles well below the angle of friction, increasing their stability.

d) (Multiple) rotational slump or rotational slide.

e) A coherent rock (basalt) rests on weaker Jurassic sediments. The Jurassic sediments are vulnerable to undercutting by weathering and erosion and, being weak, once undercut, cannot support the weight of the basalt. The slopes fail as a rotational slide.

f) A dolerite sill interrupts the weak Jurassic sediments and defines the base of the landslides. Dolerite is a highly resistant igneous rock which is not susceptible to the mass movements which affect the weaker overlying rocks.

What makes this a good answer?

Part a) The answer is precise and wholly accurate. The definition of mass movement is **clear** and the **brief examples provide just enough development**.

Part b) The answers require a brief statement followed by an explanation. Two clear statements (deforestation and loading of slopes) are made, which are accurate and **easily identified by the examiner**. Both explanations are properly focused on the question and show **appropriate application of knowledge and understanding**.

Part c) Again, both answers begin with a clear and accurate statement which the examiner can immediately identify. Two methods of stabilising slopes are described, with some development on how they operate. **The second parts of the answers are economical and wholly to the point**.

Part e) The answer provides a reasoned **explanation** of the landslides on Skye and makes excellent use of the information provided by the diagram. A common mistake in this type of question is **to describe rather than explain** the processes which have caused slope failure.

Part f) The answer makes **effective use of technical terms** such as dolerite sill, mass movements and resistant igneous rock. Again, the **evidence of the diagram is used appropriately** and the answer is accurate and **well structured**.

How to score full marks

a) **(i)** The discharge of a stream or river where the flow just reaches the top of the banks. Any further increase in discharge causes the water to spill over the banks and flood the valley floor.

(ii) The ratio of the cross-sectional area of a river channel (in square metres) to its wetted perimeter (in metres). Hydraulic radius is a measure of the efficiency of a river channel for conveying water. The higher the value of the hydraulic radius the more efficient the channel.

b) Bankfull discharge increases from 27 cumecs, at the upstream site 1, to 96 cumecs at the downstream site 2. This is because the total area of the drainage basin increases downstream and the River Aire is joined by several tributary streams, including Otterburn Beck.

c) The downstream increase in bankfull discharge can be accommodated by the River Aire in a number of ways. Apart from an increase in width and depth, there may be an increase in channel gradient or in the river's planform (e.g. the channel could become more meandering). This would explain why the increase in width and depth is less than the increase in discharge.

d) Bankfull discharge has a huge influence on channel shape because at the bankfull stage, a river has its maximum erosive and transporting power. Once bankfull is exceeded, the river loses energy as floodwaters spread across the valley floor.

e) **1** Flood embankments or levees. Flood embankments raise the sides of a river's channel and thus increase its bankfull capacity. This reduces the risk of flooding. But flood embankments also increase the height of floodwater, and if failure occurs, high-speed surges of water onto the flood plain can be devastating.

2 Dams to contain floodwaters. An entire flood can be held behind a dam and be released slowly over a long period. The main disadvantage of this flood control measure is its high capital cost. The reservoirs, impounded by dams, flood valleys and may result in the loss of farmland, settlements and amenity.

What makes this a good answer?

Part a)
(i) The answer is **accurate**, **succinct** and, in the second sentence, contains **just sufficient development** to confirm secure understanding.
(ii) The answer would probably earn full marks from the first statement. However, the **additional details** on channel efficiency and the values for the hydraulic radius are **relevant**, and suggest **excellent understanding**.

Part b) The key idea that drainage basin area increases downstream is **clearly made**. The student also provides **appropriate exemplification** from Figures 1 and 2 (discharge figures and naming a tributary stream), which ensures that the answer achieves maximum marks.

Part c) This answer is **direct** and **wholly relevant**. The student has expressed a complex idea **accurately** and **effectively**.

Part d) The **significance** of bankfull discharge for erosion and sediment transport is **clearly understood**. The reference to what happens after bankfull discharge is exceeded complements the answer, and leaves no doubt about the student's understanding.

Part e) Both answers give **a brief but clear description** and **explanation** of two flood control measures. There are **appropriate references** to the disadvantages of these measures, and just enough **exemplification** to justify full marks.

How to score full marks

a)   A beach is an accumulation of sand and shingle on the coast, between the high water and low water marks.

b)   The correct terms, reading from left to right, are:
breakpoint bar, beach face, berm and storm beach.

c)   1   Rivers. Rivers transport sand and shingle (bedload) into the coastal system. This material is then transported by wave action and tidal currents on-shore, where it accumulates as beaches.

       2   Cliff erosion. Cliff erosion results in rockfall and the input of rock particles into the coastal system.

d)   **(i)**   The beach at Blackhall is much steeper than the Cresswell beach. The Blackhall beach rises by nearly 2.5 metres over a distance of 28 metres, compared with a rise of less than 1 metre over a similar distance at Cresswell. The average gradient of both beaches is fairly constant. Neither beach has a well-developed face or berm.
The main reason for the difference in the profiles of the Blackhall and Cresswell beaches is beach material. Shingle is much more permeable than sand. When swash moves up a shingle beach rapid percolation causes it to lose energy quickly. Rapid percolation creates little or no backwash. This means that that shingle is only moved in one direction – up the beach – where it piles up at a steep angle. Sand beaches have a stronger backwash than shingle beaches, and this helps to lower beach gradients.

      (ii)   Wave type can also influence beach profiles. High energy waves create flat beaches, with well-developed breakpoint bars. Low energy waves move beach sand and shingle on-shore, and build beaches with steep faces and prominent berms.

e)   Beaches stop waves. A broad beach will absorb wave energy, and will minimise coastal erosion. Beaches adjust rapidly to changing inputs of wave energy. Thus high energy waves create broad flat beaches, which dissipate wave energy. The best defence against coastal erosion is a well developed beach.

What makes this a good answer?

Part a)   This is a **succinct and accurate** definition. One sentence is quite sufficient to achieve full marks.

Part c)   The answer **clearly states** two sources of beach material, and provides **appropriate development**.

Part d)
  (i)   This answer has a clear structure with description and explanation dealt with separately. The description makes valid comparisons between the beaches, and uses actual data from Figure 2. The explanation shows an accurate understanding of processes, and is well expressed.
  (ii)   The response contains appropriate knowledge and understanding, and is both accurate and precise.

Part e)   Another example of an **accurate response**, with **clear and precise expression** and the **use of knowledge** which is wholly **relevant**.

How to score full marks

a) The sun or solar energy.

b) Herbivore: either insects or kangaroo rats or gophers; top carnivore: coyote.

c) The roadrunner is an omnivore because it gets its food energy by eating both plant matter and other animals. It feeds directly on desert plants such as sage brush and mesquite, and eats lizards and scorpions.

d) Primary producers are green plants. They fix solar energy in a process known as photosynthesis, and convert it into sugars and starches. The primary producers occupy the first trophic level in food webs and food chains, and are the ultimate source of energy for all animals in an ecosystem.

e) The biomass (i.e. the weight of living organisms in an ecosystem) decreases at each successive trophic level in a food web or food chain. This is because animals convert only a fraction of the energy in the food they consume into living tissue. Most of the energy is lost as waste products, or as heat in respiration and other metabolic processes, which keep the animal alive.

f) Detritivores such as fungi and soil micro-organisms decompose dead plant and animal material. This is a vital function: decomposition releases essential minerals and nutrients (into the soil or atmosphere) such as calcium, magnesium, carbon and nitrogen, which can be recycled. Without this mineral cycling, ecosystems would quickly run-out of nutrients and would not survive.

g) Named ecosystem: temperate deciduous forest

Broad-leaved trees such as oak, ash and birch dominate primary production in the temperate deciduous ecosystem in the British Isles. Other primary producers include shrubs and small trees such as holly, hazel and hawthorn which form a secondary layer, and spring flowering herbs such as primroses, bluebells, violets, ransoms and dog's mercury. Primary production only occurs in the growing season between April and October. Average net primary production is relatively high at 1200 grams/m^2/year.

Primary consumers include insects (e.g. caterpillars, beetles), some seed-eating birds, and mammals such squirrels, mice and roedeer. Carnivores include birds such as owls and hawks, and mammals such as hedgehogs, stoats and foxes. Persecution by humans has caused the extinction of larger mammals such as bears, wolves, and wild boar in the British Isles in the past thousand years.

The upper layers of the soil and leaf litter host detritivores, including soil bacteria and other micro-organisms, fungi, insects and earthworms. The detritivores break down the leaf litter, releasing the nutrients into the soil, where they are absorbed by the roots of trees and recycled.

What makes this a good answer?

Part c) The student's answer demonstrates a **clear understanding** of the term 'omnivore', and further illustrates this with **specific reference** to the food web in the diagram.

Part d) This answer shows an **effective understanding of the key idea**. References to processes such as photosynthesis and the accurate use of technical terms (e.g. trophic levels, food webs, food chains etc.) confirm that this is a high quality response.

Part e) The process of energy loss through animal metabolism is clearly stated, and the answer is developed effectively by appropriate references to biomass, respiration etc. **A good balance is achieved between the general and the specific**.

Part f) The answer **focuses on the key idea** of nutrient cycling. Examples of nutrients and a recognition that nutrients cycles may include a soil or atmospheric phase, gives the answer substance.

Part g) This answer achieves the highest level of response. It has several qualities. First it is **well structured**, with separate paragraphs devoted to producers, consumers and detritivores. Second it contains **relevant detail** of plant and animal species and their niches within the temperate deciduous forest food web. Thirdly, the **knowledge presented is accurate and is applied appropriately** to the question.

5 Atmosphere

How to score full marks

a) Depression

b) A weather front is a boundary which separates air masses of contrasting temperature, humidity and lapse rate.

c) A – cold front, B – warm front, C – occluded front

d) The cloud over the North Sea and southern Scandinavia is associated with the warm and cold fronts of a depression, centred between Iceland and northern Scotland. Air rises along these fronts, creating organised bands of thick cloud, and prolonged spells of precipitation.

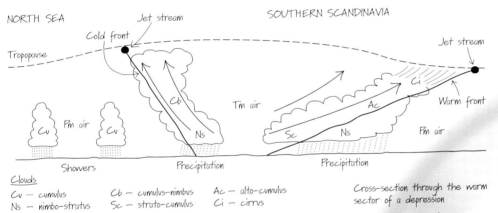

Cross-section through the warm sector of a depression

Clouds
Cu – cumulus   Cb – cumulus–nimbus   Ac – alto–cumulus
Ns – nimbo–stratus   Sc – strato–cumulus   Ci – cirrus

e) At 12.00 London occupies the warm sector of a depression. The warm sector is a wedge of tropical air and this explains London's relatively high temperature. Glasgow is situated in colder, polar air, behind the cold front of the depression. This explains why Glasgow's temperature is 3.5 degrees lower than London's.

f) Glasgow experiences a westerly wind at 12.00 on 31 March. The evidence from the weather chart (Figure 1) that explains the westerly flow is:
(a) winds blow roughly parallel to isobars, which at Glasgow are aligned east-west
(b) winds circulate anticlockwise around the depression to the north of Scotland, thus the wind must be blowing from west to east, and not east to west.

g) (i) Figure 1 shows a cold front moving south towards London at 12.00. By 15.22 the satellite image reveals a thick band of cloud across south-east England. There will be some precipitation in London. This precipitation will be prolonged, and as temperatures are well above freezing, will be in the form of rain rather than sleet or snow. The pattern of isobars indicates a moderate west to south-westerly wind.

(ii) During the next few hours the cold front will clear London and move into the North Sea and the near continent. Clouds will clear, the rain will stop and there will be an abrupt fall in temperature (three or four degrees). The night will be clear and there could be a slight ground frost. Winds will remain moderate, from a west or south-westerly direction.

What makes this a good answer?

 Part b) This is a **concise**, **accurate** and **clear definition**.

 Part d) This answer provides a **brief and accurate description** of the processes operating along frontal zones in depressions. The cross-section through the warm sector of a depression, showing both the warm and cold fronts, provides **additional detail**. The cross-section gives a **simple yet clear description** of form and process in a depression.

 Part e) The answer **concentrates only on the information relating** to weather in Figure 1 (i.e. there is no attempt to explain temperature differences in terms of latitude, ocean currents, etc.). The student demonstrates **accurate understanding** of the air masses in depressions, and the **significance** of fronts. The answer is **structured** and **clear**.

 Part f) At the outset there is a **clear statement** of wind direction. The reasoning is **fully explained**. The student provides a **helpful structure** by separating the reasons into parts (a) and (b).

 Part g)

(i) This is a Level 2 answer and is awarded full marks (see the notes on page 5 of the Introduction for guidance on level-marked questions). The answer is **detailed**, and **considers several aspects** of weather (i.e. temperature, cloud, precipitation, wind direction). There is **good use of terminology** (e.g. cold front and isobars) and the ideas are **presented logically and economically**. There is evidence that the student has used both the weather chart and the satellite image.

(ii) Again, this answer achieves full marks at Level 2. The answer **addresses the question relevantly** and shows **good understanding** of the sequence of weather changes that occur with the passage of a cold front. The student has made **effective use** of both **stimulus materials**.

6 Plate Tectonics

How to score full marks

a) Gondwanaland

b) The process of sea floor spreading explains continental drift. Volcanic activity forms new crust at mid-ocean ridges (or constructive plate margins) such as the Mid-Atlantic Ridge. The pressure exerted by this new crust and convection in the upper mantle slowly push the older oceanic crust away from the mid-ocean ridge (2 to 3 cm/year). Continents, surrounded by oceanic crust (like logs in an ice flow) are carried along by the conveyor of sea floor spreading. This is continental drift.

c) Imagine two continents that are converging, and that are separated by a narrowing sea. As the continents begin to engage, sediments on the ocean floor are squeezed, and are pushed upwards to form fold mountain ranges. This sequence of events occurred in the past 80 million years as India migrated northwards and collided with Eurasia. The Tethys Sea, separating India and Eurasia, narrowed, and the sediments on the sea floor were crumpled to form the Himalayas.

d) A – magma chamber, B – dyke, C – lava flow, D – landslide, E – crater or vent

e) Japan is located on a destructive plate boundary. The country and its adjacent seas comprise four major plates: Pacific, Philippine, Eurasian and North American. Along the east coast of Japan, the Pacific and Philippine plates are subducted beneath the Eurasia and North American plates. Subduction causes melting to occur at depth. This melt rises to the surface where it forms a chain of active volcanoes (e.g. Fuji, Unzen), stretching from Kyushu in the south to Hokkaido in the north. Most of the Japanese archipelago has formed from eruptions of ash and lava during the past few million years.

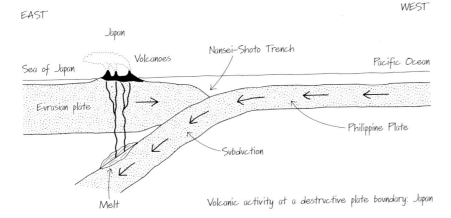

EAST   WEST

Japan

Sea of Japan   Volcanoes   Nansei-Shoto Trench   Pacific Ocean

Eurasian plate

Philippine Plate

Subduction

Melt

Volcanic activity at a destructive plate boundary: Japan

f) Tourism, energy production and agriculture all derive benefits from volcanoes. Volcanic features such as geysers and hydrothermal springs are major tourism attractions in areas such as Yellowstone (USA) and Rotorua (New Zealand). The Fuji volcano in Japan is the most visited mountain in the world. In Iceland, hot rocks close to the surface are used to generate steam and electricity, and provide central heating for the nation's capital, Reykjavik. Volcanic ash and lava weather to form fertile soils. These soils are the basis for intensive agriculture in many parts of south-east Asia (e.g. Java) and in Sicily (around the Mount Etna volcano).

What makes this a good answer?

Part b) This is a **precise answer**, addressing the question well. It demonstrates **good application of knowledge, and understanding. References to actual places** such as the Mid-Atlantic Ridge, and rates of sea floor spreading, provide the necessary detail for a Level 3 answer (see the notes on page 5 of the Introduction for guidance on level-marked questions).

Part c) This answer illustrates **a general tectonics process by referring in detail to a specific example. This is an excellent strategy** and produces an answer that is focused and wholly to the point. The answer is also accurate and clearly expressed.

Part e) Again this answer is based around a specific example. The student shows a sound grasp of processes and detail (e.g. including names of plates and volcanoes in Japan). **The cross-section is a simple, clear diagram which includes relevant labels. It is not over-elaborate**, and complements the written answer.

Part f) This answer achieves Level 3 (see the notes on page 5 of the Introduction for guidance on level-marked questions) because the student has:
- considered **several economic activities**
- made **effective connections** between these activities and volcanoes
- illustrated these with **specific examples drawn from a range of places**.

How to score full marks

a) The crude birth rate is the number of live births per 1000 of the population per year.

b) The crude birth rate is influenced by a population's age and sex structure. In a youthful population, where a large proportion of women are aged between 18 and 39, the crude birth rate will be comparatively high. This is despite the fact that on average women may only have total of one or two children each.

c) (i) The threefold increase in India's population between 1951 and 2001 was probably due to natural increase. Figure 1 shows that throughout this period the crude birth rate was much higher than the crude death rate. Figure 1 also shows that in 1951 the natural increase rate was 1.9 per cent. By 2001 it had fallen just three points to 1.6 per cent.

   (ii) Figure 1 shows that India's crude birth rate has fallen rapidly since 1971 to 24 per 1000, while the crude death rate has almost levelled out to 8 per 1000. If we project these trends into the future, then in 30 or 40 years' time India should achieve zero growth.

d) The demographic transition describes the change from high vital rates (i.e. birth rates and death rates) in a county or region, to low vital rates. At the beginning and end of this process population growth is low. However, because the death rate declines before the birth rate, the intervening period is one of rapid population growth.

e) The cause of demographic change is economic development. Rising living standards , better sanitation and health care associated with economic development lead to a fall in the death rate. The birth rate eventually falls as artificial contraception becomes widely available, and as children (who may be in full-time education for 10 years or more) are no longer an economic asset.

f) The steep fall in the crude birth rate, the levelling off of the death rate, and the decline in the rate of natural increase are similar to stages 2 and 3 of the demographic transition.

g) The demographic transition describes the population changes that occurred in western and northern Europe between 1850 and 1950. The changes were the result of industrialisation and the benefits of better diets, environmental conditions, and higher standards of living. The situation in most LEDCs today is quite different from that in Europe in the nineteenth and twentieth centuries. Many LEDCs (e.g Bangladesh) have experienced a dramatic fall in crude death rates owing to the introduction of modern medical technology. Many countries in sub-Saharan Africa show no signs of declining rates of population growth. Here children are an economic asset (they can contribute to the family income at an early age). As a result it often makes economic sense to have large families. Meanwhile there are often strong cultural and religious pressures for women to have large families, not just in Africa but also in Islamic countries in the Middle East and South Asia. Finally, even when women in LEDCs desire smaller families, lack of family planning clinics and access to contraceptives often make this impossible.

What makes this a good answer?

Part a) This answer is both **concise and accurate**.

Part b) The crucial point – that the crude birth rate is influenced by a population's age-sex structure – is clearly made at the outset. The rest of the answer provides brief, but **appropriate development of this point**. The last sentence shows that the student **understands** the difference between the crude birth rate and fertility.

Part c)

(i) This is a good answer: the student understands the significance of natural population change and **makes excellent use of the information in Figure 1** by extracting data on rates of natural increase.

(ii) The answer shows sound logic and deduction from the data in Figure 1. The birth and death trends are described accurately, **actual birth and death rate statistics are quoted** and the student's projected trends, based on the available evidence, are plausible.

Part d) This answer provides a **clear and precise definition** of the demographic transition. It shows good understanding of the concept of transition and the demographic processes operating.

Part e) This is **a relevant answer**, where **knowledge and understanding** are applied appropriately. The changes accompanying economic development are stated and, within the limits of six lines, the student gives a surprising amount of detail.

Part f) The answer **effectively** relates the experience of population change in India to the demographic transition.

Part g) This is a Level 3 answer (see the notes on page 5 of the Introduction for guidance on level-marked questions), combining both **detail and accuracy**. The student refers to three or four **valid differences** between MEDCs and LEDCs. Moreover, these differences are supported by a number of **place specific examples**.

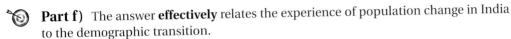

8 Migration

How to score full marks

a) Net migrational change is the difference between the number of migrants moving into an area and the number moving out. If the number of in-migrants exceeds the number of out-migrants there is a net migrational gain. An excess of out-migrants over in-migrants is a net migrational loss.

b) There was a net-migrational gain over most of southern Britain in 1998, especially in the South east, East, East Midlands, South west and Wales. In the first four regions, the net-migrational gain was over 10 000. In Wales it was lower – between 1 and 10 000. In contrast, all the regions of northern Britain suffered a net-migration loss. These losses were greatest in the North west. Two regions stand out as exceptions to this north-south divide. London had a net-migrational loss of more than 10 000 in 1998. The West Midlands region also experienced a net-migrational loss.

c) **(i)** Large numbers of retirees migrating into a region would cause an ageing of the population. With a larger proportion of people in the older age groups, the birth rate would fall, the death rate would rise and population growth would decline. This has occurred in retirement resorts in the UK such as Eastbourne and Lytham St Annes.

(ii) Large numbers of young adults leaving a region would also increase the average age of the population, lower the birth rate and increase the death rate. With a net migrational loss and a falling birth rate, the population could decline absolutely. This is known as depopulation. Large parts of upland Britain have suffered out-migration of young adults and depopulation in the past 100 years.

d) **(i)** Push factors are the adverse economic, social, political and environmental conditions in a migrant's place of origin that encourage out-migration (e.g. unemployment).

(ii) Pull factors are the economic, social, political and environmental attractions at a destination which encourage in-migration (e.g. educational opportunities).

e) The large scale migration of Mexicans to the southern USA in the past 20 years can be understood through Lee's migration model. In the early 21st century about one million Mexicans a year were making the border crossing to the USA. The majority were illegal migrants who were arrested and returned to Mexico. Migration from Mexico to the USA is primarily economic. Mexico is a relatively poor country: the USA is the world's richest. Wage rates in Mexico are barely one-tenth of those in the USA. The difference in wealth between the two countries is an enormous pull factor. Potential migrants assess positive and negative attributes of life in Mexico and in the USA. For migrants, the advantages of moving (e.g. higher standard of living, education and health care for children etc.) outweigh the disadvantages (break up of family and friends, problems of assimilation etc). However, there are obstacles to migration. These include the cost of the journey and the dangers of travelling through the desert to avoid detection at the border.

What makes this a good answer?

🎯 **Part a)** This answer is more than just a definition. Understanding of net migrational change is clear and accurate, and the answer is further developed with reference to net migrational gain and net migration loss.

🎯 **Part b)** The answer begins appropriately by describing the overall pattern of migrational change. Detail is added, with references to specific regions and rates of change. Exceptions to the overall pattern (e.g. London) are stated.

🎯 **Part c)**

(i) The focus is on retirement migration. The connection between migration and population change is explained, and the answer is rounded off with supporting examples.

(ii) The example of out-migration by young adults, causing a decline in vital rates and depopulation, is clearly explained. Again, brief supporting examples add further credibility to the answer.

🎯 **Part d)** Each term is defined accurately and concisely. To make absolutely certain of securing full marks, each answer includes a brief example.

🎯 **Part e)** This extended answer successfully relates all of the key points in Lee's model (negative and positive attributes of places, intervening obstacles and so on) to migration from Mexico to the USA. Because the answer is detailed and relevant, and is set within the context of a specific place example, it achieves Level 3 (the top level of response, see the notes on page 5 of the Introduction).

9 Rural Settlement

How to score full marks

a) From Table 1, central places with the most services are fewest in number. In Lower Wharfedale, out of the 12 services listed, 15 central places have three or fewer services. Three central places have from 4 to 6 services, Wetherby and Otley (third order) have 8 and 9 services; only Harrogate (fourth order) has all 12 services.

b) Second-order central places serve more people than first-order central places do. This means that they have the thresholds needed to support a wider range of functions than first-order central places.

c) The principal change is the declining service provision in first-order central places. Overall, first-order centres lost 14 functions between 1984 and 2001, mainly because of the closure of rural post offices. Some functions were also lost in second-order centres (3) and third-order centres (2). No settlements gained new functions.

d) 1 Commuters moving into first-order settlements. Commuters may do most or all of their shopping in nearby centres such as Leeds and Bradford where they work. This would reduce demand for village services.

2 Growth of edge-of-town superstores. These stores are easily accessible by car from surrounding rural and semi-rural areas and compete with village services.

e) (i) The average distance between centres of the same size increases with the order of the centres. Thus the average distance between the smallest centres (first-order) is from 2 to 4 kilometres. In contrast the two third-order centres (Otley and Wetherby) are approximately 20 kilometres apart.

(ii) The range of a good or service is the maximum distance people travel to purchase it. People need a low-order service such as a post office regularly. Its threshold can be met within a short distance of its location. As a result its range is small, causing the close spacing of low-order centres. Higher-order centres sell goods and services which are bought less often, and have larger thresholds and ranges. Thus these centres are more widely spaced.

f) Witherslack is a village of about 450 people, located in the south-east corner of the Lake District National Park. In the past 30 years its population has increased by the in-migration of (a) commuters to Kendal and Lancaster and (b) retirees. Despite Witherslack's population growth, demand for village services has declined. Today the only shop in the village is a small general store and sub-post office. Mobile shops no longer visit the village, and its newsagent and tobacconist has closed. The village has no public transport. Its primary school remains open, but with a roll of just 37 pupils, the school could close. A village GP has a surgery on just two days a week.

These changes have affected some residents more than others. Hardest hit are the least mobile – pensioners, poorer people without cars, young mothers at home without a car. The nearest large food store is in Kendal, 15 minutes drive time away, and the nearest centre with a full range of essential shops is Grange, 10 minutes drive time away. People must travel to obtain essential services. Those without private transport have to rely on the goodwill of friends and relatives.

What makes this a good answer?

Part a) One feature of a central place hierarchy is clearly stated. **Evidence to support a hierarchy of places** in Lower Wharfedale **is then taken from Table 1.** Excellent use is made of Table 1, with **specific data quoted**.

Part b) This is a **succinct and accurate** answer. There is clear understanding of the concept of threshold.

Part c) This is a detailed and accurate answer which achieves the highest level of response (i.e. Level 2, see the notes on page 5 of the Introduction for guidance on level-marked questions). **Information is extracted from Table 1 and is used appropriately**.

Part d) For each answer a plausible reason is stated. A **brief development** of each reason is sufficient to earn full marks.

Part e)
(i) This gives a **precise** description which focuses on the spacing of centres and provides **exemplification from Figure 1**.
(ii) The answer begins appropriately with a sound **definition** of *range*. The level of **knowledge is good**, and the **connections** between the frequency of use of services, threshold, range and the spacing of settlements are made effectively.

Part f) This answer achieves Level 3 because:
- it is linked throughout to an **actual place**, containing detail on changes in population and service provision which are **specific** to Witherslack
- it addresses **both parts of the question**
- the student's knowledge and understanding have been **applied appropriately** to the question.

How to score full marks

a) **(i)** The racial, cultural and linguistic identity of an individual or group e.g. Afro-Caribbean, white Anglo-Saxon, Muslim, etc.

(ii) The stage of an individual in the family or life cycle e.g. young adult single, young married/cohabiting without children, older married/cohabiting with grown-up children etc.

b) Some stages in the family cycle demand particular types of housing and access to services. Young adult singles may prefer apartments close to the city centre, giving good access to leisure activities and employment. Families with young children will need more space, preferably with a garden, and close to schools. Most often these requirements can be met in the suburbs.

c) Low-income populations are uniformly concentrated in Tijuana, on the Mexican side of the border. In the vast majority of census areas in Tijuana more than 30 per cent of the population have low incomes. In San Diego the distribution of the low-income population has more geographical variation. A Y-shaped low-income area focuses on the city centre, with sectors radiating south, north-west and north-east. In these sectors most areas have in excess of 40 per cent of their population in the low-income category.

d) **1** This is an inner city area. It is likely to be an area of older, cheaper housing (apartments, tenements etc.). Given the age of the housing and the possible poor quality of services and the environment, rents will be low and affordable for low-income groups.

2 This area has good access to San Diego's CBD and may attract low-income groups who find employment in jobs such as cleaning etc. in offices, shops etc. in the city centre.

e) Area B is located adjacent to the Pacific Ocean. Such a location may offer a high-quality environment with beaches and outstanding views. This would attract expensive housing developments which would exclude low-income groups.

What makes this a good answer?

Part a)
(i) The student **recognises** that ethnic identity concerns cultural and linguistic characteristics as well as racial (i.e. physical) ones.
(ii) The answer **makes references** to the main component of family status – stage in the family/life cycle – and also provides some **relevant examples**.

Part b) The answer provides **specific and accurate examples** of how family status can influence location. This **detail** is **essential** for answers to achieve the highest level of response.

Part c) Answers achieving Level 2 (see the notes on page 5 of the Introduction for guidance on level-marked questions) will be **well structured**, **detailed**, **accurate** and **use information** from Figure 1. In this answer, the main features of the spatial pattern are **clearly stated**, and the differences between Tijuana and San Diego **recognised**. There are references to directions, shapes (e.g. use of terminology such as sectors) and to percentages of low-income populations. The requirements of a Level 2 answer are fully met.

Part d) The two reasons given are quite **different** and both are **plausible**. The student shows **good understanding** of the distribution of low-class housing in cities in MEDCs and the job opportunities for low-income groups in the CBD. References to **specific details** (e.g. types of housing, the nature of jobs in the CBD) leave the examiner in no doubt that these are top-quality answers.

Part e) The answer is **clearly structured**: a **simple statement** is followed by a **brief explanation**. The student **uses** the map **evidence accurately** and the **logic** of the answer is **fully explained**.

How to score full marks

a) **(i)** hydro-electric power **(ii)** oil

b) 1 Most renewable energy resources such as solar power and hydro-electric power are inexhaustible, recyclable and will last indefinitely. In contrast non-renewable energy resources such as coal and oil are finite.

 2 Renewable energy does not pollute or degrade the environment. During combustion, non-renewable energy resources, which are mainly fossil fuels, release pollutants such CO_2 and SO_2 into the atmosphere.

c) Country A's energy relies on fossil fuels for 80 per cent of its primary energy production. Country B is less dependent on fossil fuels (70 per cent of primary energy production), and its mix of fossil fuels is very different. Natural gas dominates fossil fuels in country B, whereas oil has the dominant position in country A. Both countries produce nuclear energy, though in country A nuclear energy is nearly four times more important than in country B. Renewable energy (including HEP) contributes only 5 per cent to country A's energy economy. In country B renewable sources are much more significant. Hydro-electric power is particularly important, accounting for one-fifth of all primary energy production.

d) Country A has greater dependence on fossil fuels than country B. It will therefore make a relatively greater contribution to CO_2 pollution and global warming. Moreover, 67 per cent of energy production from fossil fuels is from coal and oil. These fossil fuels are more polluting than natural gas, which figures more prominently in country B's primary energy production. Nuclear energy is also a potential source of pollution. Country A is much more reliant on this energy source than country B. Finally, over one quarter of country B's primary energy comes from HEP and other renewables. They have few degrading effects on the environment. In comparison, renewable sources only account for 5 per cent of primary energy production in country A.

e) The main feature of France's primary energy production is its dependence on nuclear power. Seventy-five per cent of electricity in France comes from nuclear power stations. The government has encouraged the expansion of nuclear energy in the past 30 years because France has limited reserves of coal and oil. Natural gas is an important source of primary energy production thanks to the country's own gas fields in the south-west and imported gas from nearby North Africa.

 Hydro-electric power is significant, accounting for around 13 per cent of France's electricity production. France has large resources of HEP in the Alps and on major rivers such as the Rhône and Durance.

 Oil is essential for road transport. However, the absence of indigenous oil reserves in France has encouraged the government to minimise oil imports by substituting other fuels for oil (e.g. nuclear) wherever possible.

 Thus France has a distinctive energy economy, strongly influenced by its lack of domestic oil and coal reserves. It depends more heavily on nuclear power than almost any other MEDC, while hydro power makes a major contribution to electricity production.

What makes this a good answer?

Part b) Both answers are **clear**, **concise** and **focused**. There is **accurate use of terminology** such as 'recyclable', 'fossil fuels', 'pollutants', which shows **good understanding**.

Part c) The answer is **well structured**, **summative** and **provides the necessary detail**. The comparison is made point-by-point, and is a single coherent answer rather than separate descriptions of two countries. There is a **appropriate and specific reference to figures** from the two pie charts.

Part d) The connection between fossil fuels and their environmental impact is **appreciated** and is **clearly stated** at the outset. The answer is **effectively** developed and **explains** the causal link between fossil fuel consumption and atmospheric pollution, and the less damaging effects of renewables. Overall, this answer shows a **high level of understanding**.

Part e) The answer is **wholly place specific**. It is based around one country – France – and the amount of **detail provided** (e.g. energy consumption, energy mix, resources etc.) and its **relevant application** to the question, are impressive.

12 Hypothesis Testing 1

How to score full marks

a)   Rock type influences slope steepness in the Wenlock Edge area.

b)   **(i)**   Two point locations – one on the Aymestry Ridge and the other in Ape Dale – would be selected, using randomly generated northings and eastings. These two points would be joined to form a line of transect. Along the line of transect 50 slope sections, each 5 metres in length, would be measured at 20 metre intervals on each of the five major rock types.

   (ii)   The purpose of this transect sampling is to collect data which will accurately represent slopes on the five rock types. This means selecting data objectively. The transect line is selected randomly, and the 5 metre slope sections are at 20 metre intervals to ensure coverage over the largest area. If the slope sections were not separated in this way, the transect on each rock type would be just 250 metres. Over such a short distance, small scale features such as river valleys could distort the results. Spatial sampling using transects also has the advantage of minimising the time spent collecting data in the field.

c)   **(i)**   The data could be represented as histograms. Histograms are a type of bar chart which show the frequency distribution of a single variable. In this example slope angles would be divided into six or seven classes (e.g. 0–2.9, 3–5.9, 6–8.9, 9–11.9, 12–14.9, 15–17.9, 18 and above). The number of slope sections in each class would then be plotted as a histogram.

   (ii)   The slope angles plotted as histograms are unlikely to follow a symmetrical or normal frequency distribution (i.e. they will be skewed). Thus the mean or average is unlikely to represent the data accurately. The median slope angle for each of the five rock types would give a more accurate picture of how slopes vary with rock type. This is because the median is simply the middle value in a data set arranged in rank order. The median is not affected by extreme values in the distribution.

What makes this a good answer?

Part a)   This hypothesis is a statement which is **clear**, **plausible** and **testable**.

Part b)   The answer describes an objective sampling method based on a line transect chosen at random. This method is **practicable** and will generate a fairly large data base in a relatively short time. The answer is **clear, concise and an accurate response** to the command word 'describe'.

Part c)

   (i) Histograms are the most appropriate method for representing the data. The answer shows **sound understanding** of this type of chart and explains in some detail how data from the investigation could be adapted to a histogram format. Other charts, such as pie charts and divided (stacked) bar charts, although perhaps less effective than histograms, would also be acceptable.

   (ii) The answer shows a **critical awareness** of the mean and median. The important **link** between central tendency measures and the nature of the frequency distribution is clearly made. Thus the logic of using the median rather than the mean is irrefutable.

How to score full marks

a) The distance travelled by commuters increases with income status (or type of housing occupied). Commuters with higher incomes (i.e. those resident in higher value housing) probably travel further than those with lower incomes. Two possible reasons are:

1 Those on higher incomes can afford to spend more on commuting.

2 Higher-income jobs are less likely to be available locally than lower-income jobs.

b) **(i)** The accuracy of a sample increases with the square of its size (e.g. to increase accuracy twofold requires a sample 2^2 or 4 times larger). Other things being equal, larger samples are likely to provide more precise and accurate data than smaller samples.

(ii) All of the street/road names on the map could be listed in alphabetical order. If there were 30 streets/roads, ten could be selected systematically e.g. first, third, sixth, ninth … .From each of the ten streets/roads, ten houses would be selected at regular intervals (e.g. if there are 50 houses, every fifth house).

(iii) Systematic sampling is simple and can be done quickly. Once a random interval for selecting the sample has been chosen no further preparation is required. Systematic sampling is objective, and provided there are no underlying regularities in the population (e.g. a street of terraces where every nth house is detached) it should provide an accurate sample. However, with a small sample, systematic sampling methods may fail to pick out sub-groups in a population. For instance, if the housing in Figure 1 were 50 per cent terrace, 25 per cent semi-detached and 25 per cent detached (corresponding to low-, medium-, and high-income groups) there is no guarantee that a systematic sample would represent it in these proportions. As a result we might have just four or five households from detached houses in our sample. This sample would be too small to test the hypothesis meaningfully.

c) Stratified systematic sampling is an alternative to systematic sampling. We use stratified sampling where the population comprises a number of sub-groups. In this example the houses in the suburbs could be divided into three types: terrace, semi-detached, detached. Systematic sampling could then be used to select 33 houses from each sub-group. With three samples of equal size we could then compare distances travelled to work for each group. Such a comparison might involve drawing histograms, calculating mean values and even using statistical analysis such as chi-squared.

What makes this a good answer?

Part a) A geographical idea, derived from the information on the map, is presented as a **credible hypothesis** that should be **testable**.

Part b)

(i) The key idea – the larger the sample the greater the accuracy – is correct and **clearly stated**. The detail (level of accuracy increases with the square of sample size) is not essential for full marks, but helps to make this a very good answer.

(ii) A concise answer which conveys a **secure knowledge and understanding** of systematic sampling. **Exemplification** (e.g. sampling interval) **clarifies the student's understanding**. These qualities put the answer at the top end of Level 2 (see the notes on page 5 of the Introduction for guidance on level-marked questions).

(iii) The answer deals separately with the strengths and weaknesses of systematic sampling. **The question does not require a comparison and so this treatment is appropriate**. Descriptions of strengths and weaknesses show accurate knowledge and critical understanding of sampling methods. The **descriptions are also linked explicitly to the hypothesis under investigation**.

Part c) A Level 2 answer would contain **accurate knowledge and understanding** of an alternative sampling method, and a **reasoned justification** for its use. This answer meets both criteria. There is a detailed and accurate description of stratified sampling, with a brief but valid explanation of its use **in the context of the investigation**.

How to score full marks

a) A dot map is a type of statistical map which represents the distribution of a given number of people (in this example) by a dot of constant size.

b) **(i)** Dot value will depend on the number of people in the parishes and the scale of the map. If the dot value is too large (e.g. 500) then some parishes will contain no dots at all (e.g. Gunthorpe) and the distribution would look very thin. Too small a value is just as bad. A value of ten people per dot would mean 578 dots in Sheringham and the map would have too many dots to show any clear pattern. The dot value chosen has to avoid these two extremes.

   (ii) Dots are not placed randomly within each parish. The distribution of the main settlements in each parish (from a 1:50 000 or 1:100 000 map) should guide dot placement. An effort should also be made to place dots in such a way as to ensure a continuous distribution of dots across parish boundaries.

c) A dot map shows the actual location of the population across an area. The dots are placed within the parishes to correspond with the main areas of settlement. A dot map therefore has a precision which most other statistical maps lack.

d) **1** Choropleth maps are based on areas (e.g. the area of each parish in north Norfolk). The number of people in each parish is to some extent influenced by the size of the parish (i.e. larger parishes usually contain more people than smaller parishes). Mapping the number of people in each parish would not therefore provide an accurate view of population distribution in north Norfolk.

   2 Choropleth maps tell us nothing about the distribution of population within the parishes. This is a particular problem if there are large parishes where most people are concentrated in just one settlement (e.g. Holt). In these circumstances the result is a generalised and misleading distribution.

e) The data in Table 1 could also be represented as a proportional symbol map. Circles or squares with areas proportional to population could be drawn in each parish. This type of map usually shows absolute values within areas. It would therefore be an appropriate technique for mapping the populations of each parish. Moreover the proportional symbols would allow an estimate of the population of each parish.

What makes this a good answer?

🎯 **Part a)** This concise and **accurate** definition requires no further elaboration.

🎯 **Part b)**
   (i) Identifying the two main influences on dot value (population of parishes and mapping scale) is worth 2 marks. These points are developed, with **appropriate reference to parishes in north Norfolk**. This **accurate, relevant answer** gains full marks at Level 2 (see the notes on page 5 of the Introduction for guidance on level-marked questions).
   (ii) Stating and recognising that dots must be located according to the distribution of settlement in each parish and the need for a large-scale map shows both **good understanding** and an ability to **apply knowledge appropriately**. This is a Level 2 answer.

🎯 **Part c)** One advantage of dot maps is stated and is awarded 1 mark. There is enough development of this initial statement to merit a further 2 marks.

🎯 **Part d)** Two disadvantages of choropleth maps are recognised, then **placed within the context of the population data** for north Norfolk, making the answer Level 2. Answers only specified two **general** weaknesses of choropleth maps would be Level 1.

🎯 **Part e)** Proportional symbol maps are correctly identified as another way of mapping population distribution. The justification for using this method is credible and shows **sound understanding** of mapping techniques. The answer gains the full 3 marks.